Letts

gets you through

ITEM

D0256485

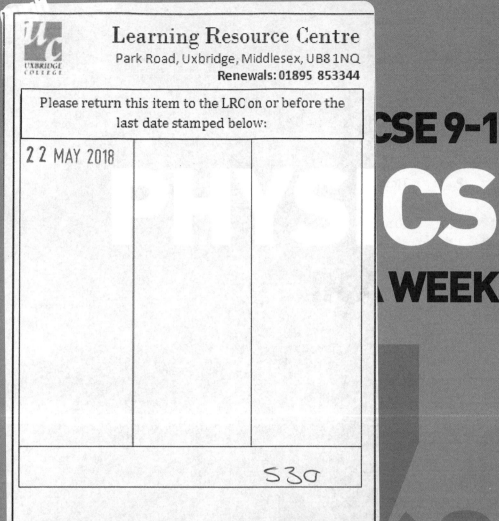

UXBRIDGE COLLEGE

Learning Resource Centre

Park Road, Uxbridge, Middlesex, UB8 1NQ

Renewals: 01895 853344

Please return this item to the LRC on or before the
last date stamped below:

2 2 MAY 2018

S3o

GCSE 9-1
PHYSICS
IN A WEEK

7 DAYS

DAN FOULDER

Revision Planner

HT Upthrust

A submerged object experiences a greater pressure on the bottom surface than on the top surface. This creates a resultant force upwards. This is **upthrust**.

An object less dense than the surrounding liquid displaces a volume of liquid equal to its own weight. This object floats as its weight is equal to the upthrust.

An object more dense than the surrounding liquid is unable to displace a volume of liquid equal to its own weight. This object sinks as its weight is greater than the upthrust.

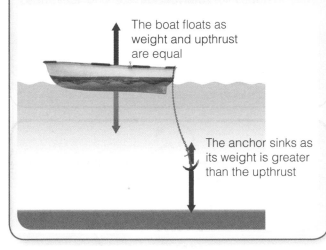

The boat floats as weight and upthrust are equal

The anchor sinks as its weight is greater than the upthrust

Atmospheric pressure

The atmosphere is a thin layer (relative to the size of the Earth) of air round the Earth. The atmosphere gets less dense with increasing altitude. The air molecules colliding with a surface create atmospheric pressure.

As height of an object increases, there are fewer air molecules above the object so their total weight is smaller. This leads to a lower atmospheric pressure.

SUMMARY

- The pressure in a fluid is caused by the fluid and the surrounding atmospheric pressure.
- Liquids and gases are fluids.
- HT Upthrust is the resultant force when a submerged object experiences greater pressure on the bottom surface than the top.
- The thin layer of air around the Earth is called the atmosphere.
- As the height of an object increases, the atmospheric pressure decreases.

QUESTIONS

QUICK TEST

1. Explain why air pressure decreases with height.
2. HT What is the pressure in a 19 m column of water? (density of water = 1000 kg/m^3 and gravitational field strength = 10 N/kg)
3. HT What is upthrust?

EXAM PRACTICE

1. A hydraulic system uses liquids to transmit pressure and magnify a force.

   a) Explain why liquids are used in hydraulic systems instead of gases. **[2 marks]**

   b) In a hydraulic system, what force is exerted when the pressure is 1200 Pa and the area of the surface is 0.2m^2 ? **[2 marks]**

2. HT A student carried out an investigation into floating and sinking of model boats using different materials.

   a) Her first boat sank. Explain, in terms of forces, why this occurred. **[2 marks]**

   b) How could the student modify her boat to increase the chance of it floating? **[2 marks]**

HT indicates content that is Higher Tier only.

WS indicates 'Working Scientifically' content, which covers practical skills and data-related concepts.

Pressure in a fluid

Fluids

Liquids and gases are **fluids**.

The pressure in a fluid is caused by the fluid and the surrounding atmospheric pressure. The pressure causes a force to act at right angles (normal) to a surface. The pressure exerted on a surface by a fluid can be calculated using the following equation:

$$pressure = \frac{\text{force normal to a surface}}{\text{area of that surface}}$$
$$p = \frac{F}{A}$$

- pressure, p, in pascals, Pa
- force, F, in newtons, N
- area, A, in metres squared, m^2

Example

What is the pressure of a force of 650 N over an area of 0.50 m^2?

$$pressure = \frac{\text{force normal to a surface}}{\text{area of that surface}}$$
$$= \frac{650}{0.5}$$
$$= 1300 \text{ Pa}$$

Liquids are relatively incompressible and the pressure in liquid is transmitted equally in all directions, meaning hydraulic systems containing liquids can be used to transmit pressure and so magnify a force.

HT The pressure exerted by a column of liquid can be calculated using the following equation:

$$pressure = \begin{array}{c}\text{height} \\ \text{of the} \\ \text{column}\end{array} \times \begin{array}{c}\text{density} \\ \text{of the} \\ \text{liquid}\end{array} \times \begin{array}{c}\text{gravitational} \\ \text{field} \\ \text{strength}\end{array}$$

$$p = h\rho g$$

- pressure, p, in pascals, Pa
- height of the column, h, in metres, m
- density, ρ, in kilograms per metre cubed, kg/m^3
- gravitational field strength, g, in newtons per kilogram, N/kg

Example

What is the pressure at a depth of 40 m in seawater? (density of seawater = 1029 kg/m^3 and gravitational field strength = 10 N/kg)

pressure = $40 \times 1029 \times 10$

= 411.6 kPa

Pressure increases in a column of liquid as there are more particles applying the pressure. If the height of the column increases or the density of the liquid increases, this causes an increase in the number of particles and so increases the pressure.

Newton's laws

Newton's first law

Newton's first law deals with the effect of resultant forces.

- If the resultant force acting on an object is zero and the object is stationary, the object remains stationary.
- If the resultant force acting on an object is zero and the object is moving, the object continues to move at the same speed and in the same direction (its velocity will stay the same).

Newton's first law means that the velocity of an object will only change if a resultant force is acting on the object.

Examples of resultant force:

A car is stationary. At this point the resultant force is zero.

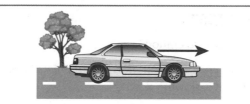

The car starts to move and accelerates. The car is accelerating as the resultant force on it is no longer zero.

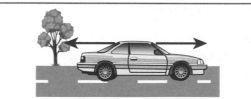

The car travels at a constant velocity. The resultant force is zero again.

HT The tendency of objects to continue in their state of rest or of uniform motion is called inertia.

Stationary object – zero resultant force

Object moving at constant speed – zero resultant force

Newton's second law

The acceleration of an object is proportional to the resultant force acting on the object, and inversely proportional to the mass of the object.

Therefore:

acceleration ∝ resultant force

resultant force = mass × acceleration

- force, F, in newtons, N
- mass, m, in kilograms, kg
- acceleration, a, in metres per second squared, m/s^2

Example

A motorbike and rider of mass 270 kg accelerate at 6.7 m/s^2. What is the resultant force on the motorbike?

$$270 \times 6.7 = 1809 \text{ N}$$

If the motorbike slows to a constant speed, the resultant force would now be 0 (as acceleration = 0, $270 \times 0 = 0$).

HT **Inertial mass** is a measure of how difficult it is to change the velocity of an object. It is defined by the ratio of force over acceleration.

Newton's third law

Whenever two objects interact, the forces they exert on each other are equal and opposite.

When a fish swims it exerts a force on the water, pushing it backwards. The water exerts an equal and opposite force on the fish, pushing it forwards.

SUMMARY

- Newton's first law says that the velocity of an object will only change if a resultant force is acting on the object.
- Newton's second law says that the acceleration of an object is proportional to the resultant force acting on it, and inversely proportional to its mass.
- Newton's third law says that when two objects interact, the forces they exert on each other are equal and opposite.

QUESTIONS

QUICK TEST

1. A sprinter of mass 93 kg accelerates at 10 m/s^2. What is the resultant force?

2. A book of weight 83 N is on a desk. What force is the desk exerting on the book?

3. What happens to the speed of a moving object that has a resultant force of 0 acting on it?

4. What happens to the speed of a stationary object that has a resultant force of 0 acting on it?

EXAM PRACTICE

1. A cannonball with a mass of 7kg when fired from a cannon experiences a force of 2100N.

   a) What is the acceleration of the cannonball? **[2 marks]**

   b) What force does the cannon experience?

      Explain how you arrived at your answer. **[3 marks]**

ⓗⓣ Momentum

Momentum

Moving objects have momentum. Momentum is given by the following equation:

momentum = mass × velocity

$$p = mv$$

- momentum, p, in kilograms metre per second, kg m/s
- mass, m, in kilograms, kg
- velocity, v, in metres per second, m/s

Example

A car of mass 1100 kg is travelling at 13 m/s. What is its momentum?

$p = mv$

$\quad = 1100 \times 13$

$\quad = 14\,300$ kg m/s

Conservation of momentum

In a closed system, the total momentum before an event is equal to the total momentum after the event. This is called **conservation of momentum**.

Example

A cue ball with mass of 0.17 kg travelling at 4 m/s hits a stationary snooker ball with a mass of 0.16 kg. The first ball stops while the other ball starts to move. What is the velocity of the second ball?

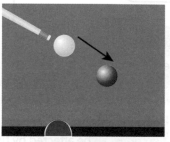

Before collision

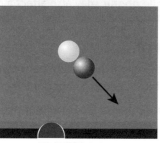

After collision

momentum = mass × velocity

momentum of first ball
before collision = 0.17 × 4 = 0.68 kg m/s

momentum of first ball after collision = 0 (as it is now stationary)

As momentum is the same before and after the collision, the second ball has a momentum of 0.68 kg m/s.

$\quad 0.68 = $ mass × velocity

$\quad 0.68 = 0.16 \times$ velocity

$\dfrac{0.68}{0.16} = $ velocity

velocity = 4.25 m/s

Changes in momentum

When a force acts on an object that is moving or is able to move, a change in momentum occurs.

The equations $F = m \times a$ and

$$a = \frac{v - u}{t}$$

lead to the equation

$$F = \frac{m\Delta v}{\Delta t}$$

- $m\Delta v$ = change in momentum

Example

When a 20 g dart travelling at 18 m/s hits a dartboard it comes to rest in 0.2 s. What force is generated by this impact?

initial momentum of dart = mass × velocity

$$= 18 \times 0.02$$

$$= 0.36 \text{ kg m/s}$$

final momentum = 0 kg m/s

change in momentum = 0.36 − 0

$$= 0.36$$

$$\text{force} = \frac{\text{change in momentum}}{\text{time}}$$

$$= \frac{0.36}{0.2}$$

$$= 1.8 \text{ N}$$

Safety features

Safety features include:

- airbags
- seat belts
- gymnasium crash mats
- cycle helmets
- cushioned surfaces for playgrounds.

All these features slow down the change in momentum in a collision. This reduces the forces on the people involved, reducing injury.

SUMMARY

- All moving objects have momentum.
- Conservation of momentum means the total momentum before an event is equal to the total momentum after the event.

QUESTIONS

QUICK TEST

1. Name two car safety features that slow down the change in momentum in a collision.

2. How do airbags reduce the chance of injury in a collision?

3. What is the momentum of an object of 8 kg moving at 5 m/s?

EXAM PRACTICE

1. A car is travelling at 9m/s when it hits a crash barrier at the side of the road. The driver has a mass of 75 kg. As he is wearing a seat belt he comes to rest in 0.1 seconds.

   a) What is the force produced by the driver's change in momentum? **[3 marks]**

   b) How would the force have been affected by the following factors?

      In each case, explain your answer.

      i) The driver not wearing a seat belt. **[2 marks]**

      ii) The driver wearing a seat belt and the car being fitted with an airbag. **[2 marks]**

Changes in energy

A system is an object or group of objects. When a system changes, so does the way energy is stored within its objects.

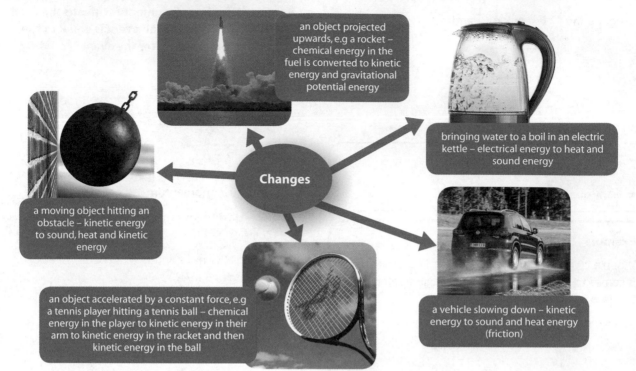

an object projected upwards, e.g a rocket – chemical energy in the fuel is converted to kinetic energy and gravitational potential energy

bringing water to a boil in an electric kettle – electrical energy to heat and sound energy

Changes

a moving object hitting an obstacle – kinetic energy to sound, heat and kinetic energy

an object accelerated by a constant force, e.g a tennis player hitting a tennis ball – chemical energy in the player to kinetic energy in their arm to kinetic energy in the racket and then kinetic energy in the ball

a vehicle slowing down -- kinetic energy to sound and heat energy (friction)

Energy in moving objects

The kinetic energy of a moving object can be calculated using the following equation:

kinetic energy = 0.5 × mass × (speed)²

$$E_k = \frac{1}{2}mv^2$$

- kinetic energy, E_k, in joules, J
- mass, m, in kilograms, kg
- speed, v, in metres per second, m/s

Example

A ball of mass 0.3 kg falls at 10 m/s. What is the ball's kinetic energy?

$$E_k = \frac{1}{2}mv^2$$
$$= 0.5 \times 0.3 \times 10^2$$
$$= 15 \text{ J}$$

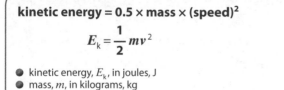

Elastic potential energy (Ee)

elastic potential energy = 0.5 × **spring constant** × **extension²**

$$E_k = \frac{1}{2}mv^2$$

(assuming the limit of proportionality has not been exceeded)

- elastic potential energy, E_e, in joules, J
- spring constant, k, in newtons per metre, N/m
- extension, e, in metres, m

Example

A spring has a spring constant of 500 N/m and is extended by 0.2 m. What is the elastic potential energy in the spring?

$$E_e = \frac{1}{2}ke^2$$
$$= \frac{1}{2}500 \times 0.2^2$$
$$= 250 \times 0.04$$
$$= 10 \text{ J}$$

Gravitational potential energy (g.p.e.)

The amount of gravitational potential energy gained by an object raised above ground level can be calculated using the following equation:

g.p.e. = mass × gravitational field strength × height

$$E_P = mgh$$

- gravitational potential energy, E_p, in joules, J
- mass, m, in kilograms, kg
- gravitational field strength, g, in newtons per kilogram, N/kg
- height, h, in metres, m

Example

A car of mass 1500 kg drives up a hill which is 67 m high. What is the gravitational potential energy of the car at the top of the hill? (Assume the gravitational field strength is 10 N/kg)

$$E_p = mgh$$
$$= 15\,000 \times 67 \times 10$$
$$= 10\,050 \text{ kJ}$$

Changes in thermal energy

The specific heat capacity of a substance is the amount of energy required to raise the temperature of one kilogram of the substance by one degree Celsius. The specific heat capacity can be used to calculate the amount of energy stored in or released from a system as its temperature changes. This can be calculated using the following equation:

change in thermal energy = mass × specific heat capacity × temperature change

$$\Delta E = m\,c\,\Delta\Theta$$

- change in thermal energy, ΔE, in joules, J
- mass, m, in kilograms, kg
- specific heat capacity, c, in joules per kilogram per degree Celsius, J/kg °C
- temperature change, $\Delta\Theta$, in degrees Celsius, °C

Example

0.75 kg of 100°C water cools to 23°C. What is the change in thermal energy?

(Specific heat capacity of water = 4184 J/kg °C)

$$\Delta E = mc\,\Delta\Theta$$
$$= 0.75 \times 4184 \times (100 - 23)$$
$$= 241\,626 \text{ J} = 242 \text{ kJ}$$

SUMMARY

- Kinetic energy can be calculated using the equation: ½ x mass x (speed)^2
- Gravitational potential energy can be calculated using the equation: mass x gravitational field strength x height
- Elastic potential energy can be calculated using the equation: 0.5 x spring constant x extension^2
- The specific heat capacity of a substance is the amount of energy needed to raise the temperature of 1 kg by 1°C.

QUESTIONS

QUICK TEST

1. What is specific heat capacity?

2. What is the kinetic energy of a 620 g object travelling at 5 m/s?

3. What is the gravitational potential energy of a 17 kg object at a height of 456 m? (gravitational field strength = 10 N/kg). Give your answer to 4 significant figures.

EXAM PRACTICE

1. A slingshot is used to fire a ball diagonally upwards. The slingshot has a spring constant of 115 N/m and it is extended by 0.49m.

   a) Calculate the elastic potential energy of the slingshot. Give your answer to 3 significant figures. **[2 marks]**

   b) What energy transfer takes place when the slingshot is fired? **[2 marks]**

   c) The ball lands on the ground. What is its gravitational potential energy at this point?

   Explain your answer. **[2 marks]**

Conservation and dissipation of energy

Energy transfers in a system

Energy can be **transferred**, **stored** or **dissipated**. It cannot be created or destroyed.

In a closed system there is no net change to the total energy when energy is transferred.

Only part of the energy is usefully transferred. The rest of the energy dissipates and is transferred in less useful ways, often as heat or sound energy. Energy is then described as **wasted**.

Examples of processes which cause a rise in temperature and so waste energy as heat include:

- friction between the moving parts of a machine
- electrical work against the resistance of connecting wires.

If the energy that is wasted can be reduced, that means more energy can be usefully transferred. The less energy wasted, the more efficient the transfer.

Efficiency

The energy efficiency for any energy transfer can be calculated using the following equation:

$$\text{efficiency} = \frac{\text{useful output energy transfer}}{\text{total input energy transfer}}$$

Efficiency may also be calculated using the following equation:

$$\text{efficiency} = \frac{\text{useful power}}{\text{total power output}}$$

Example

Kat uses a hairdryer. Some of the energy is wasted as sound. The electrical energy input is 24 kJ. The energy wasted is 7 kJ. What is the efficiency of the hairdryer?

First calculate the useful energy transferred.

useful energy = total energy – wasted energy

$$= 24 - 7 = 17 \text{ kJ}$$

Now calculate the efficiency.

$$\text{efficiency} = \frac{\text{useful output energy transfer}}{\text{total input energy transfer}}$$

$$= \frac{17}{24} = 0.71$$

This decimal efficiency can be represented as a percentage by multiplying it by 100.

$$0.71 \times 100 = 71\%$$

HT There are many ways energy efficiency can be increased:

- Lubrication, thermal insulation and low resistance wires reduce energy waste and improve efficiency.
- Thermal insulation, such as loft insulation, reduces heat loss.
- Low resistance wires reduce energy lost as heat when an electrical current flows through them.

Power

One of the definitions of power is work done over time (the rate at which work is done). The power equation is:

$$\text{power} = \frac{\text{work done}}{\text{time}} \qquad P = \frac{W}{T}$$

- P = power (watts)
- E = work done (joules)
- T = time (seconds)

Example

A weightlifter is lifting weights that have a mass of 80 kg. What power is required to lift them 2 metres vertically in 4 seconds? (Assume gravitational field strength of 10 N/kg)

$$\text{power} = \frac{\text{work done}}{\text{time}}$$

$$\text{work done} = \text{force} \times \text{distance}$$

$$\text{force} = \text{mass} \times \text{gravitational field strength}$$

$$= 80 \times 10 = 800 \text{ N}$$

$$\text{work done} = 800 \times 2 = 1600 \text{ J}$$

$$\text{power} = \frac{1600}{4}$$

$$= 400 \text{ W}$$

A second weightlifter lifted the same mass to the same height in 3 seconds. As he carried out the same amount of work but in a shorter time, he would have a greater power.

Thermal insulation

Thermal insulation has a low thermal conductivity, so has a slow rate of energy transfer by conduction. U-values give a measure of the heat loss through a substance. A higher U-value indicates that a material has a higher thermal conductivity.

Changing the material of walls to materials that have a lower thermal conductivity reduces heat loss (the U-value) and so a building cools more slowly, reducing heating costs. The graph below shows the heat lost by different types of wall.

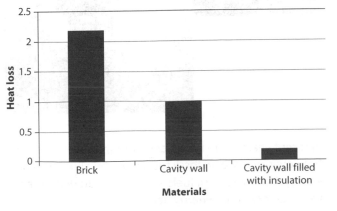

The house below is not fitted with insulation, so lots of heat is being lost through the walls due to their high thermal conductivity. This can be seen by the red/orange colour on the infrared image.

SUMMARY

- Energy can be transferred, stored or dissipated.
- Energy cannot be created or destroyed.
- Power means work done over time.
- Thermal insulation has a slow rate of energy transfer by conduction.

QUESTIONS

QUICK TEST

1. What is thermal insulation?

2. What is the power of a model crane which does 50 J of work in 2 seconds?

3. What effect does fitting cavity wall insulation have on the thermal conductivity of the building's wall?

EXAM PRACTICE

1. A washing machine transfers 600 J of useful energy out of a total of 802 J.

   What is the efficiency of the washing machine? **[1 mark]**

Transverse and longitudinal waves

Transverse waves

In a transverse wave, the oscillations are perpendicular to the direction of energy transfer, such as the ripples on the surface of water.

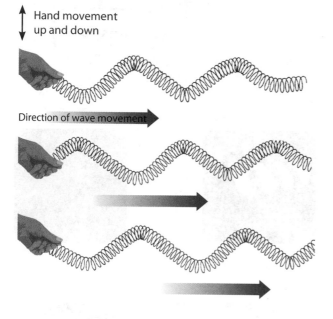

Longitudinal waves

In a longitudinal wave, the oscillations are parallel to the direction of energy transfer. Longitudinal waves show areas of compression and rarefaction, such as sound waves travelling through air.

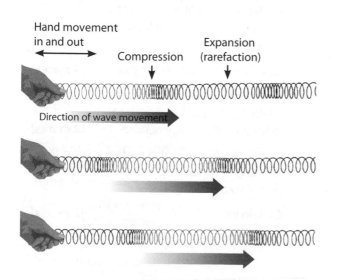

Wave movement

In both sound waves in air and ripples on the water surface, it is the wave that moves forward rather than the air or water molecules. The waves transfer energy and information without transferring matter.

This can be shown experimentally.

For example, when a tuning fork is used to create a sound wave that moves out from the fork, the air particles don't move away from the fork. (This would create a vacuum around the tuning fork.)

Properties of waves

Waves are described by their:

- **amplitude** – The amplitude of a wave is the maximum displacement of a point on a wave away from its undisturbed position.

- **wavelength** – The wavelength of a wave is the distance from a point on one wave to the equivalent point on the adjacent wave.

- **frequency** – The frequency of a wave is the number of waves passing a point each second.

- **period** – The time for one complete wave to pass a fixed point. The equation for the time period (T) of a wave is given by the following equation:

$$period = \frac{1}{frequency}$$
$$period = \frac{1}{f}$$

- period, T, in seconds, s

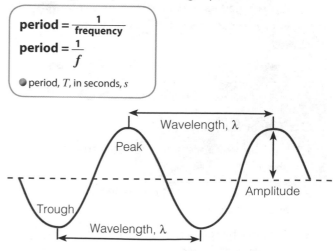

Wave speed

The wave speed or wave velocity is the speed at which the energy is transferred (or the wave moves) through the medium.

The wave speed is given by the following wave equation:

wave speed = frequency × wavelength
$$v = f\lambda$$

- wave speed, v, in metres per second, m/s
- frequency, f, in hertz, Hz
- wavelength, λ, in metres, m

Example

A sound wave in air has a frequency of 250 Hz and a wavelength of 1.32 m. What is the speed of the sound wave?

wave speed = frequency × wavelength

$$= 250 \times 1.32$$

$$= 330 \text{ m/s}$$

A ripple tank and a stroboscope can be used to measure the speed of ripples on the surface of water.

As wave speed, frequency and wavelength are all interrelated, changes in wave speed due to waves travelling between different media will affect the wavelength of a sound. For example, when sound waves enter a more dense medium, the wave speed increases. This leads to an increase in the sound's wavelength.

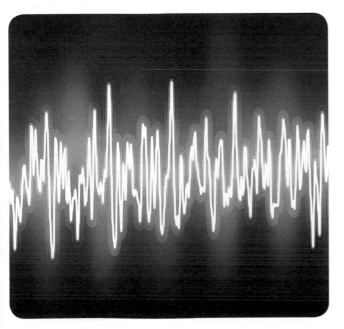

SUMMARY

- Waves can be either transverse or longitudinal.
- Waves are described by their amplitude, wavelength, frequency and period.
- Wave speed can be calculated using the equation, wave speed = frequency x wavelength.

QUESTIONS

QUICK TEST

1. What is meant by the term 'wavelength'?
2. Give an example of a transverse wave.
3. Give an example of a longitudinal wave.
4. What are the four properties waves are described by?

EXAM PRACTICE

1. A group of students are investigating waves using a ripple tank and a stroboscope.

   a) What type of waves are they investigating? **[1 mark]**

   b) They record 2 waves passing a point in 1 second.

      Calculate the period of the wave. **[2 marks]**

   c) The wavelength of the waves is 30cm.

      Calculate the speed of the wave. **[2 marks]**

   d) The students placed a small piece of cork in the tank. They observed it moving up and down as the wave passed but it did not move along the tank.

      Explain the property of waves that this provides evidence of. **[2 marks]**

HT Waves for detection and exploration

Ultrasound

Ultrasound waves have a frequency higher than the upper limit of hearing for humans.

Ultrasound waves are partially reflected when they meet a boundary between two different media. The time taken for the reflections to reach a detector can be used to determine how far away such a boundary is. This allows ultrasound waves to be used for both medical and industrial imaging.

One of the major uses of ultrasound is prenatal scans.

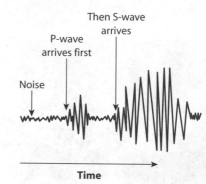

The P-waves and S-waves produced by an earthquake can be detected by a seismograph

P-waves and S-waves have been used to provide evidence for the structure and size of the Earth's core.

As it is impossible to see the internal structure of the Earth, we must use models to study it.

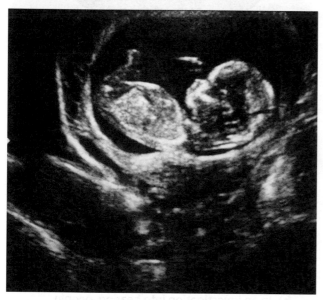

Earthquakes

Earthquakes produce two types of seismic wave that travel through the Earth:

1. P-waves are longitudinal, seismic waves. P-waves travel at different speeds through solids and liquids.

2. S-waves are transverse, seismic waves. S-waves cannot travel through a liquid.

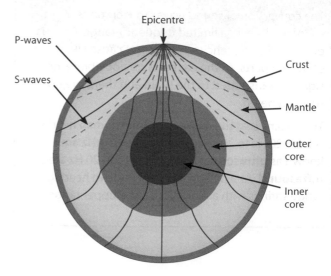

P-waves and S-waves are detected on the Earth's surface. Only P-waves are detected on the opposite side of the Earth from the epicentre of the earthquake, not S-waves. This helps prove that the outer core of the Earth is liquid, preventing the passage of S-waves.

Echo sounding

Echo sounding (sonar) using high frequency sound waves is used to detect objects in deep water and to measure water depth.

Fishing vessels use echo sounding to detect shoals of fish. The high frequency sound bounces off the group of fish and is detected by the boat. The time interval between the ultrasound being produced and detected allows the distance to the fish shoal to be determined.

Example

A fishing boat releases high frequency sound waves and detects the reflection from a shoal of fish 0.5 seconds later. The speed of sound in water is 1500 m/s. How far below the boat are the fish?

distance = speed × time

$= 1500 \times 0.5$

$= 750$ metres

$= \dfrac{750}{2}$

$= 375$ metres

This value then needs to be divided by 2 as this is how far the sound waves must travel to the fish shoal and then back to the boat.

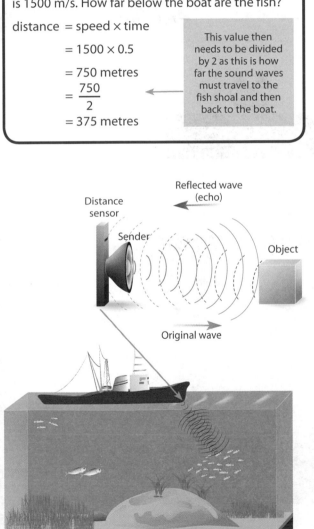

Reflected wave (echo)

Distance sensor

Sender

Object

Original wave

SUMMARY

- Ultrasound waves can be used for industrial and medical imaging, for example, prenatal scans.
- Echo sounding uses high frequency sound waves to measure water depth or detect objects in deep water.
- Earthquakes produce two types of seismic wave: longitudinal P-waves and transverse S-waves.

QUESTIONS

QUICK TEST

1. Describe two differences between P-waves and S-waves.

2. What is ultrasound?

3. What type of wave is used in echo sounding?

EXAM PRACTICE

1. After an earthquake in Chile (a country in South America) P-waves are detected in China, which is on the opposite side of the Earth's surface. S-waves are detected in countries across South America but not in China.

   Explain these observations. **[4 marks]**

2. During an ocean surveying mission, echo sounding is used to determine the depth of the sea floor.

   The reflected sound wave is detected 1.3 seconds after being released, and the speed of sound in water is 1500 m/s.

   How deep is the water? **[2 marks]**

Electromagnetic waves and properties 1

Electromagnetic waves

Electromagnetic waves are transverse waves that:

- transfer energy from the source of the waves to an absorber
- form a continuous spectrum
- travel at the same velocity through a vacuum (space) or air.

The waves that form the electromagnetic spectrum are grouped in terms of their wavelength and their frequency.

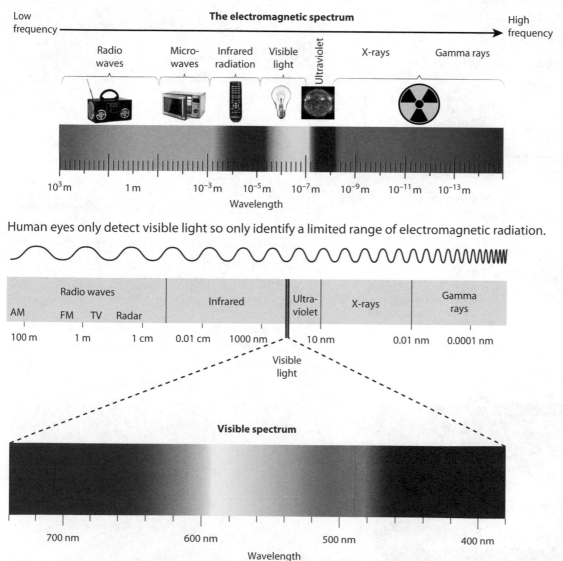

Human eyes only detect visible light so only identify a limited range of electromagnetic radiation.

Different wavelengths of electromagnetic waves are reflected, refracted, absorbed or transmitted differently by different substances and types of surface.

Refraction

Refraction is when a wave changes direction as it travels from one medium to another.

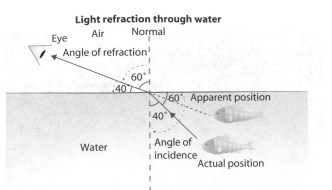

Light refraction through water

In the example above, the fish appears to be above its actual position due to the refraction of the light rays.

HT Refraction occurs due to the wave changing speed as it moves between media. Waves have different velocities in different media.

Light travels faster in the air than it does in the water. This leads to the light ray bending away from the normal. When light slows down, it bends towards the normal.

The wave front diagram below shows a wave moving from air (less optically dense) to water (more optically dense). As the wave travels slower in a denser medium, the edge of the wave that hits the water first slows down whilst the rest of the wave continues at the same speed. This causes the light to bend towards the normal. The opposite effect occurs when a wave moves from a more optically dense medium to a less optically dense medium.

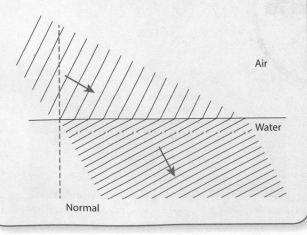

SUMMARY

- Electromagnetic waves are transverse waves.
- The waves that form the electromagnetic spectrum are ordered according to their frequency and wavelength.
- Human eyes can only detect visible light.
- When a wave changes direction as it travels from one medium to another, it is called refraction.

QUESTIONS

QUICK TEST

1. What type of electromagnetic waves are used in television remote controls?
2. Which has the highest frequency: visible light or microwaves?
3. Are electromagnetic waves transverse or longitudinal?
4. What is refraction?
5. **HT** Which way do waves bend when they enter a more optically dense medium?

EXAM PRACTICE

1. Visible light from the Sun takes the same time to reach the Earth as ultraviolet radiation from the Sun.

   Explain this observation. **[2 marks]**

2. **HT** Spear fishing is a traditional means of fishing. The fisherman stands in the water and uses a sharp stick to spear the fish as they swim by.

   Fully explain why the fishermen do not aim the sharp stick exactly where they observe the fish to be. **[3 marks]**

Electromagnetic waves and properties 2

Electromagnetic waves

Changes in atoms and the nuclei of atoms can result in electromagnetic waves being generated or absorbed over a wide frequency range.

Gamma rays originate from changes in the nucleus of an atom.

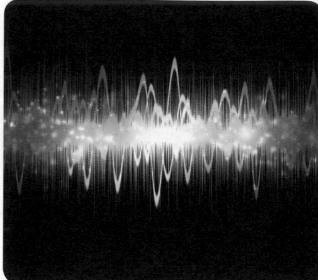

> ### HT Radio waves
>
> Radio waves can be produced by oscillations in electrical circuits. When radio waves are absorbed they may create an alternating current with the same frequency as the radio wave. Therefore, radio waves induce oscillations in an electrical circuit.

The diagram below shows the main features of electromagnetic waves.

Ultraviolet waves, x-rays and gamma rays can have hazardous effects on human body tissue.

The effects of electromagnetic waves depend on the type of radiation and the size of the dose.

Radiation dose (in sieverts) is a measure of the risk of harm from an exposure of the body to the radiation.

1000 millisieverts (mSv) equals 1 sievert (Sv)

Microwaves can cause internal heating of body cells.

Ultraviolet waves can cause skin to age prematurely and increase the risk of skin cancer.

X-rays and gamma rays are ionising radiation that can cause the mutation of genes and cancer.

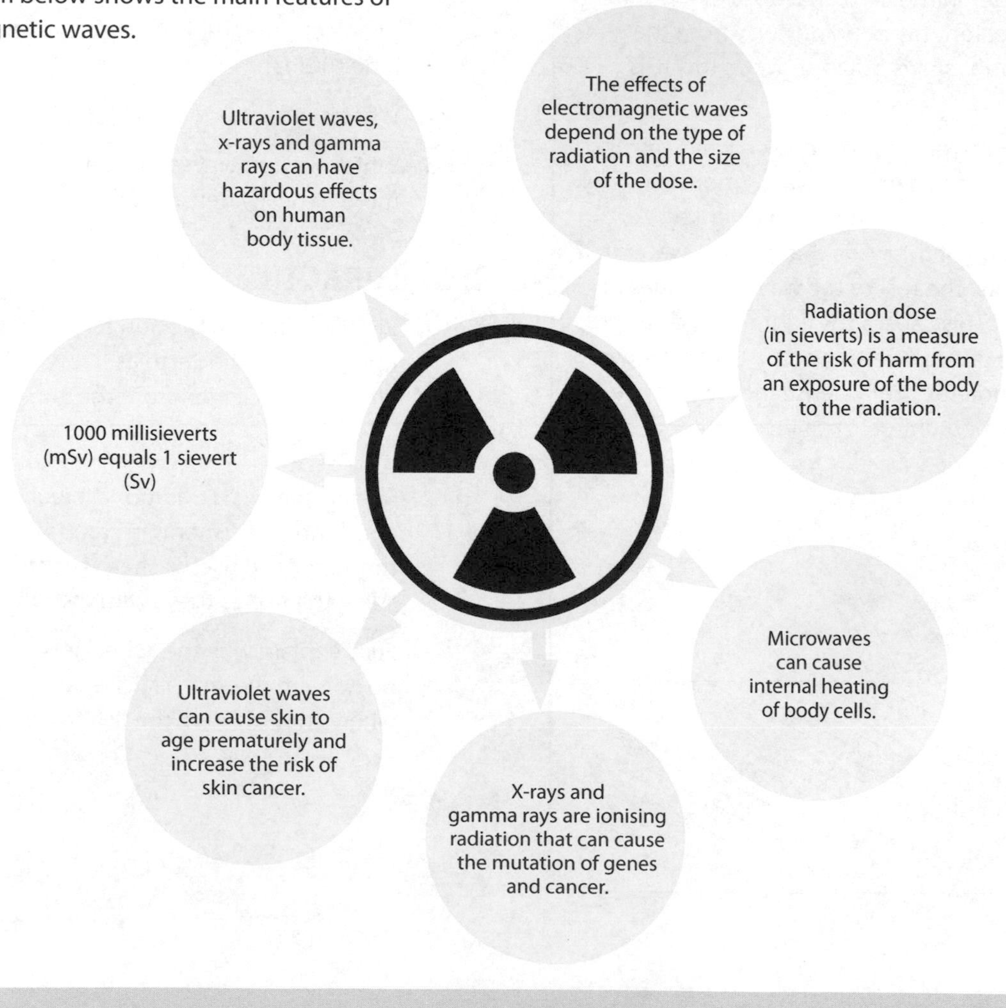

Uses of electromagnetic waves

Electromagnetic waves have many practical applications.

Radio waves	Television, radio, bluetooth	Low frequency radio waves can diffract around hills and reflect off the ionosphere, meaning they don't require a direct line of sight between transmitter and receiver. (HT)
Microwaves	Satellite communications, cooking food	Their small wavelength allows them to be directed in narrow beams. (HT)
Infrared	Electrical heaters, cooking food, infrared cameras	Thermal radiation heats up objects. (HT)
Visible light	Fibre optic communications	This can be reflected down a fibre optic cable. (HT)
Ultraviolet	Energy efficient lamps, sun tanning	Ultraviolet is used for low energy light bulbs to produce white light. This requires less energy than filament bulbs. (HT)
X-rays	Medical imaging and treatments	X-rays are absorbed differently by different parts of the body; more are absorbed by hard tissues, such as bone, and less are absorbed by soft tissues. This allows images of the inside of the body to be created. (HT)
Gamma rays	Sterilising, medical imaging and treatment of cancer	Gamma rays destroy living cells so can be used to sterilise medical equipment and apparatus. Gamma rays can also be used to destroy cancerous tumours and carry out functional organ scans. (HT)

SUMMARY

● Electromagnetic waves have many uses, for example radio waves in TV and radio, microwaves for cooking food, and X-rays and gamma rays for medical imaging.

QUESTIONS

QUICK TEST

1. How many millisieverts are in 1 sievert?
(HT) 2. How are radio waves produced?
3. What are the uses of gamma rays?
4. Give one use of infrared radiation.

EXAM PRACTICE

(HT) 1. Explain how X-rays are used in medical imaging.   [3 marks]

2. What are the dangers associated with too much exposure to ultraviolet radiation?   [2 marks]

Lenses and visible light

Lenses

A lens forms an image by refracting light.

- ● Convex lens – parallel rays of light are brought to a focus at the principal focus. Convex lenses can form virtual or real images.

- ● Concave lens – parallel rays of light diverge when they pass through a concave lens. Concave lenses always form virtual images.

Real images and virtual images

The distance from the lens to the principal focus is called the focal length.

Distance of object from convex lens	Image
More than 2 × focal length	Real and diminished
2 × focal length	Real and same size as object
Between 2 × focal length and the focal length	Real and magnified
Closer than focal length	Virtual

The effect of convex and concave lenses can be shown by ray diagrams. The lenses will be represented by symbols.

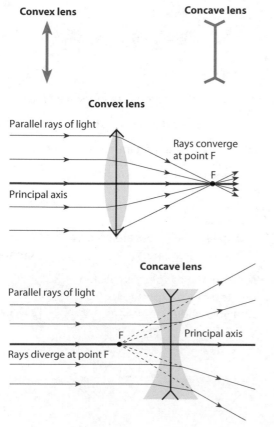

More powerful lenses have a shorter focal length. Thicker convex lenses are generally more powerful than thinner ones. The magnification produced by a lens can be calculated using the following equation:

$$\text{magnification} = \frac{\text{image height}}{\text{object height}}$$

Image heights are usually measured in mm or cm. Magnification does not have any units.

Example

What is the magnification of an image that is 8 cm high when the original object was 0.3 cm high?

$$\text{magnification} = \frac{\text{image height}}{\text{object height}}$$
$$= \frac{8}{0.3}$$
$$= 27 \times \text{magnification}$$

Visible light

Each colour within the visible light spectrum has a narrow range of wavelengths and frequency.

IR					UV
λ [nm]	780	700	600	500	380

Opaque objects reflect or absorb all light incident on them. The colour of an opaque object is determined by which wavelengths of light are more strongly reflected. Wavelengths not reflected are absorbed.

For example, a blue object absorbs all wavelengths of visible light except blue. The blue wavelength is diffusely reflected into the eyes of the observer.

A white object reflects all wavelengths of light, so the object appears white.

White object

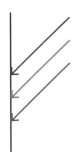

Black object

A black object absorbs
all wavelengths of light
(no reflection occurs).

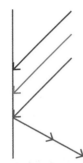

Blue object

A blue object absorbs all
wavelengths of light except
blue which it reflects. It
therefore appears blue.

There are two types of reflection:

- Specular reflection – reflection from a smooth surface in a single direction, e.g. reflection from a mirror.

- Diffuse reflection – reflection from a rough surface causes scattering. Diffuse reflection allows us to see objects.

Objects that transmit light are either transparent (allow all light to pass through) or translucent (allow light through but diffuses it, causing things observed through a translucent object to not be properly visible).

Colour filters are transparent objects that transmit certain wavelengths of light and absorb all other wavelengths of light. For example, a red filter transmits red light and absorbs all other visible light wavelengths.

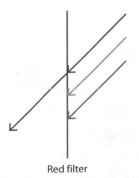

Red filter

SUMMARY

- Lenses form images by refracting light.
- There are two types of lens: convex and concave.
- The distance from the lens to the principal focus is called the focal length.
- There are two types of reflection: specular reflection and diffuse reflection.
- Objects that transmit light are either transparent or translucent.

QUESTIONS

QUICK TEST

1. What is an opaque object?

2. What type of image is formed by a concave lens?

3. What is the original height of an object that has been magnified 8 times to produce a 22 cm high image?

EXAM PRACTICE

1. a) In an investigation into coloured light, filters were used which allowed light to pass through without diffusing it.

    Were these filters transparent or translucent?

    Explain your answer.   **[2 marks]**

    b) What colour would a blue object appear when viewed through a perfectly red filter?

    Explain your answer.   **[2 marks]**

    c) The reflection of light from the blue object was an example of diffuse reflection.

    Explain how this differs from specular reflection.   **[2 marks]**

Black body radiation

Emission and absorption of infrared radiation

All bodies emit and absorb infrared radiation. The hotter the body, the more infrared radiation it emits in a given time.

A perfect black body is a theoretical object that:

- absorbs all of the radiation incident on it
- does not reflect or transmit any radiation
- is the best possible emitter as it emits the maximum amount of radiation possible at a given temperature.

A perfect black body is impossible in the real world as no object can fulfil these criteria.

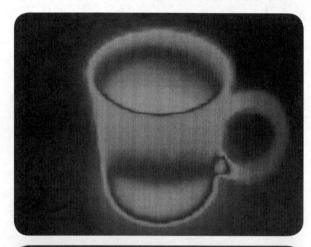

HT All bodies emit radiation, and the intensity and wavelength distribution of any emission depends on the temperature of the body.

Objects with a high temperature appear white hot, like this molten metal.

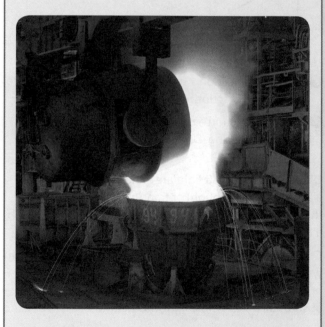

A body at a constant temperature is absorbing radiation at the same rate as it is emitting radiation.

The temperature of a body increases when the body absorbs radiation faster than it emits radiation.

The temperature of the Earth depends on many factors, as the following diagram shows, including:

- the rates of absorption of radiation
- the rate of emission of radiation
- the rate of reflection of radiation into space.

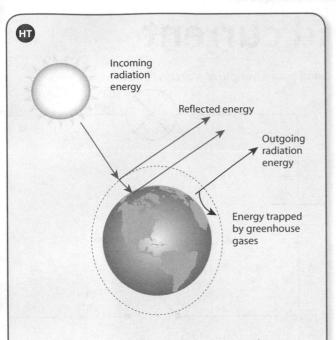

Incoming radiation energy

Reflected energy

Outgoing radiation energy

Energy trapped by greenhouse gases

The temperature of a body is related to the radiation absorbed and the radiation emitted.

For example, when food is beneath a grill it is absorbing more radiation than it is emitting so its temperature is increased. When the food is removed from the grill, it now emits more radiation than it is absorbing. This means the food cools down.

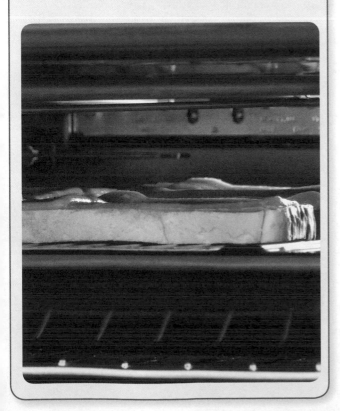

SUMMARY

● All bodies emit infrared radiation.

● All bodies absorb infrared radiation.

● A perfect black body is a theoretical object that is not possible in reality.

● **HT** The temperature of a body is related to the radiation absorbed and emitted; a body at a constant temperature absorbs and emits radiation at the same rate.

QUESTIONS

QUICK TEST

1. Describe the three main characteristics of a perfect black body.

2. Is it possible for a perfect black body to exist?

3. What will happen to the temperature of a body that is emitting as much radiation as it is absorbing?

EXAM PRACTICE

1. An investigation was carried out into the effect of heating on a metal.

    a) Use the ideas of absorbing and emitting radiation to explain the following:

        i) The metal's temperature rose when it was heated. **[1 mark]**

        ii) The metal's temperature fell when it was no longer heated. **[1 mark]**

        iii) After the metal reached room temperature its temperature remained constant. **[1 mark]**

    b) At what point in the investigation would the metal emit white light? **[1 mark]**

Circuits, charge and current

Circuit symbols

The diagram below shows the standard symbols used for components in a circuit.

Here is an example of a circuit diagram:

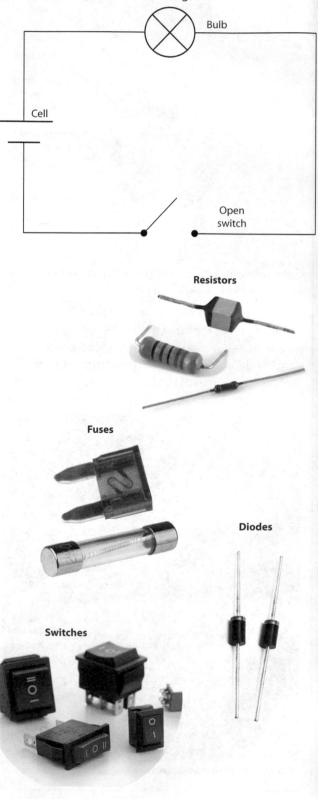

Bulb

Cell

Open switch

WS

Cell		Bulb	
Battery		Diode	
Switch (open)			
Switch (closed)		LED	
Voltmeter	**V**		
LDR		Thermistor	
		Resistor	
Variable resistor		Ammeter	**A**
Motor	**M**	Fuse	

Resistors

Fuses

Diodes

Switches

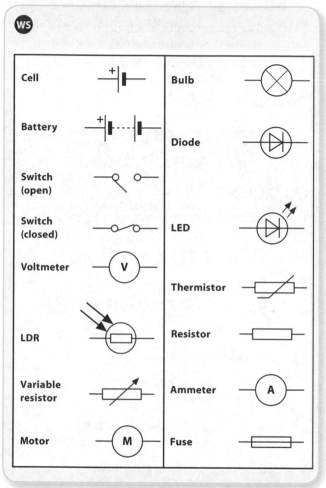

Electrical charge and current

For electrical charge to flow through a closed circuit, the circuit must include a source of energy that produces a potential difference, such as a battery, cell or powerpack.

Electric current is a flow of electrical charge. The size of the electric current is the rate of flow of electrical charge. Charge flow, current and time are linked by the following equation:

> **charge flow = current × time**
>
> $$Q = It$$
>
> - charge flow, Q, in coulombs, C
> - current, I, in amperes or amps A
> - time, t, in seconds, s

> **Example**
>
> A current of 6 A flows through a circuit for 14 seconds. What is the charge flow?
>
> charge flow = current × time
>
> $\qquad = 6 \times 14$
>
> $\qquad = 84\ C$

The current at any point in a single closed loop of a circuit has the same value as the current at any other point in the same closed loop.

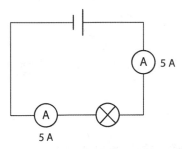

Both ammeters in the circuit above show the same current of 5 amps.

SUMMARY

- Circuit diagrams are drawn using the standard symbols for components in a circuit.
- A source of energy must be present in a closed circuit in order for an electrical charge to flow through the circuit.
- Electrical current is a flow of electrical charge.

QUESTIONS

QUICK TEST

1. What is the symbol for a resistor?

2. What charge flows through a circuit per second if the current is 3.8 A?

3. What is the current of a charge flow of 600 C in three seconds?

EXAM PRACTICE

1. **a)** A circuit with a single closed loop was being set up.

   For the electrical charge to flow in the circuit, what component must be included in the circuit?

   Explain your answer. **[2 marks]**

   b) i) The current through a diode in the circuit was 8A and the charge flow was 160 C.

   How long was the current flowing for? **[2 marks]**

   ii) What would the current be through a fuse in the same circuit?

   Explain your answer. **[2 marks]**

Current, resistance and potential difference

Current, resistance and potential difference

The current through a component depends on both the resistance of the component and the potential difference (p.d.) across the component. Potential difference is the energy transferred per unit charge passed.

The greater the resistance of the component, the smaller the current for a given potential difference across the component.

Current, potential difference or resistance can be calculated using the following equation:

> **potential difference = current × resistance**
>
> $$V = IR$$
>
> ● potential difference, V, in volts, V
> ● current, I, in amperes or amps, A
> ● resistance, R, in ohms, Ω

> **Example**
>
> A 5 ohm resistor has a current of 2 A flowing through it. What is the potential difference across the resistor?
>
> potential difference = current × resistance
> $$= 5 \times 2$$
> $$= 10 \text{ V}$$

By measuring the current through, and potential difference across a component, it's possible to calculate the resistance of a component.

The circuit diagram (right) would allow you to determine the resistance of the filament lamp.

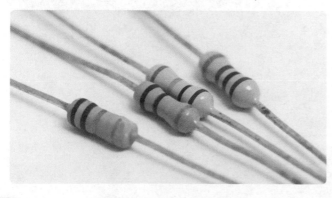

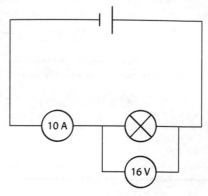

potential difference = current × resistance

resistance $= \dfrac{\text{potential difference}}{\text{current}}$

$= \dfrac{16}{10}$

= 1.6 ohms

Resistors

In an ohmic conductor, at a constant temperature the current is directly proportional to the potential difference across the resistor. This means that the resistance remains constant as the current changes.

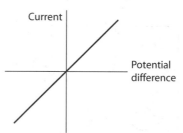

The resistance of components such as lamps, diodes, thermistors and LDRs is not constant; it changes with the current through the component. They are not ohmic conductors.

When a current flows through a resistor, the energy transfer causes the resistor to heat up. This is due to collisions between electrons and the ions in the lattice of the resistor.

This heating can be an advantage, such as in an electrical heater. It is also a disadvantage as it can lead to electrical devices being damaged due to overheating. Thicker wires have a lower resistance as there is a larger cross-sectional area for the current to pass through.

Filament lamps

The resistance of a filament lamp increases as the temperature of the filament increases.

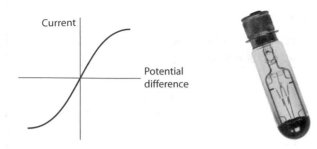

Diodes

The current through a diode flows in one direction only. This means the diode has a very high resistance in the reverse direction.

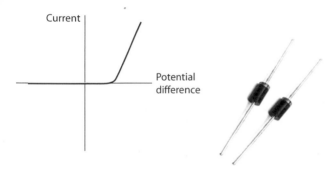

Light dependent resistors (LDR)

The resistance of an LDR decreases as light intensity increases. LDRs are used in circuits where lights are required to switch on when it gets dark, such as floodlights.

Thermistors

The resistance of a thermistor decreases as the temperature increases. Thermistors are used in thermostats to control heating systems.

SUMMARY

● Current, resistance and potential difference are related in that the current through a component depends on both the resistance of the component, and the potential difference across the component.

● Current, potential difference or resistance can be calculated using V = IR.

● In an ohmic conductor, at a constant temperature, the resistance remains constant as the current changes.

QUESTIONS

QUICK TEST

1. Why is an LDR not an ohmic conductor?

2. What is the potential difference if the current is 6 A and the resistance is 3 ohms?

3. Calculate the resistance of a component that has a current of 2 A flowing through it and a potential difference of 8 V.

EXAM PRACTICE

1. An investigation was carried out into the resistance of different components in a circuit which contained a 12V battery pack.

    a) What two pieces of equipment need to be wired into a circuit in order to determine the resistance of a component in the circuit? [2 marks]

    b) A filament lamp is used that has a resistance of 4 ohms.

       What is the current flowing through the filament lamp? [2 marks]

    c) After being left on for a period of time, the resistance of the lamp changed.

       Explain why this occurred and predict how the resistance changed. [2 marks]

Series and parallel circuits

Components can be joined together in either a series circuit or a parallel circuit. Some circuits can include both series and parallel sections.

Series circuits

For components connected in series:

● there is the same current through each component

● the total potential difference of the power supply is shared between the components

● the total resistance of two components is the sum of the resistance of each component.

Total resistance is given by the following equation:

$$R \text{ total} = R1 \times R2$$

● resistance, R, in ohms, Ω

Example

What is the total resistance of the two resistors in the series circuit below?

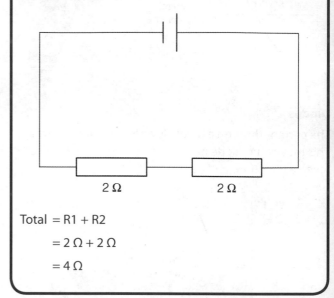

$$2\,\Omega \qquad 2\,\Omega$$

Total $= R1 + R2$

$= 2\,\Omega + 2\,\Omega$

$= 4\,\Omega$

Parallel circuits

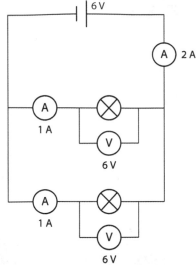

For components connected in parallel:

- the potential difference across each component is the same

- the total current through the whole circuit is the sum of the currents through the separate components. The current splits between the branches of the circuit and combines when the branches meet

- the total resistance of two resistors is less than the resistance of the smallest individual resistor. This is due to the potential difference across the resistors being the same but the current splitting.

SUMMARY

- Components can be joined together in a series circuit or in a parallel circuit.
- In a series circuit, the current is the same through each component; in a parallel circuit, the current through the whole circuit is the total of the currents through the separate components.

QUESTIONS

QUICK TEST

1. How do you calculate the total resistance of two resistors in series?

2. If the current through one component in a series circuit is 4 A, what is the current through the rest of the components?

3. What is a parallel circuit?

EXAM PRACTICE

1. In a parallel circuit a component has a potential difference across it of 9V.

   a) What would be the voltage through the other components in the circuit?

      Explain your answer.   **[2 marks]**

   b) Each component has a resistance of 3 ohms.

      What conclusion can be made about the total resistance of all the components in the circuit?

      Explain your answer.   **[2 marks]**

Domestic uses and safety

Direct and alternating current

Cells and batteries supply current that always passes in the same direction. This is direct current (dc).

Alternating current (ac) changes direction at a frequency of fifty times a second. Mains electricity is an ac supply. In the UK it has a frequency of 50 Hz and is about 230 V.

Mains electricity

> **WS** Most electrical appliances are connected to the mains using a three-core cable with a three-pin plug.
>
> | Live wire | Brown | Carries the alternating potential difference from the supply. |
> |---|---|---|
> | Neutral wire | Blue | Completes the circuit. The neutral wire is at, or close to, earth potential (0 V). |
> | Earth wire | Green and yellow stripes | The earth wire is at 0 V. It only carries a current if there is a fault. |

The potential difference between the live wire and earth (0 V) is about 230 V.

Our bodies are at earth potential (0 V). Touching a live wire produces a large potential difference across our body. This causes a current to flow through our body, resulting in an electric shock that could cause serious injury or death.

Insulation, fuses and circuit breakers

If an electrical fault causes too great a current, the circuit is disconnected by a fuse or a circuit breaker connected to the live wire.

The current will cause the fuse to overheat and melt or the circuit breaker to switch off (trip). A circuit breaker operates much faster than a fuse and can be reset.

Appliances with metal cases are usually earthed. If a fault occurs, a large current flows from the live wire to earth. This melts the fuse and disconnects the live wire.

Some appliances are double insulated meaning it is impossible for the case to become live. (Either the case is plastic or it is impossible for the live wire to come into contact with the casing). Double insulated appliances have no earth connection.

Electric drills are examples of appliances that are double insulated.

Circuit breaker

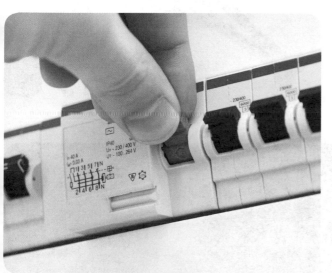

Fuses

SUMMARY

- Current that always flows in the same direction is called direct current (dc).
- Current that changes direction is called alternating current (ac). Mains electricity is ac.
- Most electrical appliances use a plug to connect to mains electricity. A plug has a live wire, neutral wire and earth wire.
- A fuse or circuit breaker disconnects the circuit if the current is too great.

QUESTIONS

QUICK TEST

1. What do ac and dc stand for?

2. What are the advantages of a circuit breaker over a conventional fuse?

3. What colour is the live wire in a plug?

4. What voltage does the earth wire have in a plug?

EXAM PRACTICE

1. A student was investigating the plugs of different electrical appliances.

   He was examining the plug of an electric appliance which had a plastic case. He noticed that this plug did not have an earth wire.

   He wrote in his report that all appliances must have an earth wire and this particular appliance was therefore unsafe to use.

   a) Was the student correct?

      Explain your answer. **[4 marks]**

   b) Explain why a fuse is important in an appliance with an earth wire. **[2 marks]**

Energy transfers

Power

The power of a device is related to the potential difference across it and the current through it by the following equations:

> **power = potential difference × current**
>
> $$P = VI$$
> **or**
> **power = current² × resistance**
>
> $$P = I^2 R$$
>
> - power, P, in watts, W
> - potential difference, V, in volts, V
> - current, I, in amperes or amps, A
> - resistance, R, in ohms, Ω

Example

A bulb has a potential difference of 240 V and a current flowing through it of 0.6 A. What is the power of the bulb?

power = potential difference × current

$$= 240 \times 0.6$$
$$= 144 \text{ W}$$

Energy transfers in everyday appliances

Everyday electrical appliances are designed to bring about energy transfers.

The amount of energy an appliance transfers depends on how long the appliance is switched on for and the power of the appliance.

Here are some examples of everyday energy transfer in appliances:

- A hairdryer transfers electrical energy from the ac mains to kinetic energy (in an electric motor to drive a fan) and heat energy (in a heating element).

- A torch transfers electrical energy from batteries into light energy from a bulb.

Work done

Work is done when charge flows in a circuit.

The amount of energy transferred by electrical work can be calculated using the following equation:

> **energy transferred = power × time**
> $$E = Pt$$
> **and**
> **energy transferred = charge flow × potential difference**
> $$E = QV$$
>
> - energy transferred, E, in joules, J
> - power, P, in watts, W
> - time, t, in seconds, s,
> - charge flow, Q, in coulombs, C
> - potential difference, V, in volts, V

The National Grid

The National Grid is a system of cables and transformers linking power stations to consumers.

Electrical power is transferred from power stations to consumers using the National Grid.

Step-up transformers **increase** the potential difference from the power station to the transmission cables.

Step-down transformers **decrease** the potential difference to a much lower and safer level for domestic use.

Increasing the potential difference reduces the current so reduces the energy loss due to heating in the transmission cables. Reducing the loss of energy through heat makes the transfer of energy much more efficient. Also, the wires would glow and be more likely to break over time if the current through them was high.

SUMMARY

- Power can be calculated using the equation: power = potential difference × current or power = current² × resistance.
- Electrical appliances bring about energy transfers, for example, a torch transfers electrical energy into light energy.
- The National Grid is a system for getting electricity to consumers. Step-up transformers increase potential difference from the power station to the cables; step-down transformers decrease potential difference to a safe level for homes.

QUESTIONS

QUICK TEST

1. Explain why step-down transformers are important in the National Grid.

2. What is the energy transferred by a charge flow of 50 C and a potential difference of 10 V?

3. What is the power of a device that has a current flowing through it of 4 A and a resistance of 3 Ω?

EXAM PRACTICE

1. An overhead cable has a current of 500 A and a power of 8000 kW.

   a) What is the resistance of the cable? **[2 marks]**

   b) What is energy transferred by the cables in 120 seconds?

   Give your answer in kJ. **[2 marks]**

   c) Explain the advantage of keeping the current relatively low in these wires. **[2 marks]**

Static electricity

Static charge

When some insulating materials are rubbed against each other, they become electrically charged. One material gains electrons from the other and becomes negatively charged. The other material loses electrons so becomes positively charged. As the charge remains on the materials, this is static electricity.

These materials now have become electrically charged. This is a static charge.

If two objects that carry the same type of charge are brought together, they repel. For example, two positively charged objects will repel.

Two objects that carry different types of charge will attract. For example, a positively charged object will attract a negatively charged object.

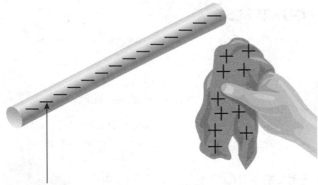

Electron

Attraction and repulsion between two charged objects are examples of non-contact forces.

The greater the charge on an isolated object, the greater the potential difference between the object and earth.

If this potential difference becomes high enough, a spark may jump across the gap between the object and an earthed conductor brought near to it.

Examples of static electricity include:

- electric shocks from everyday objects
- lightning
- a charged balloon attached to a wall
- a charged comb picking up small pieces of paper.

Earthing can be used to remove excess charge by allowing the electrons to flow down the earthed conductor. This is important in preventing dangers of sparking, such as when fuelling cars where a spark could lead to fuel igniting.

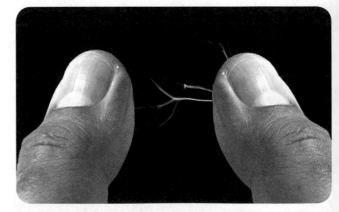

Electric fields

A charged object creates an electric field around itself. An electrical field is a region where an electrical charge experiences a force. The electric field is strongest close to the charged object. The further away from the charged object, the weaker the field. The number and density of field lines show the strength of the field – the more lines, the stronger the field.

The electric field from an isolated positive charge

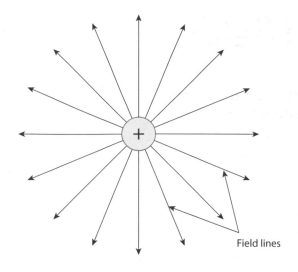

Field lines

The electric field from an isolated negative charge

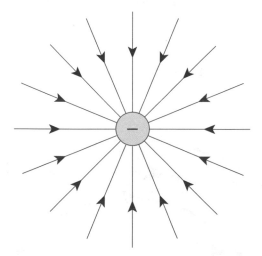

A second charged object placed in the field experiences a force. The force gets stronger as the distance between the objects decreases.

This force is the attraction and repulsion felt by charged objects when they are brought together. The field lines show the direction the force would act on a positively charged object. The electrical field is also what leads to the sparking between charged objects.

SUMMARY

● Static electricity is produced when some insulating materials are rubbed against each other.
● Two objects that carry the same charge repel when brought together; two objects that carry opposite charges attract when brought together.
● A charged object creates an electric field around itself.

QUESTIONS

QUICK TEST

1. What is static electricity?

2. What causes an object to become positively charged?

3. What happens when two objects with like charges are brought together?

EXAM PRACTICE

1. A student carried out an investigation into static electricity using a negatively charged object and a positively charged object.

   a) The student brought the two objects into close proximity with each other.

      Explain, using the concept of electrical fields, the forces the two objects would experience. **[3 marks]**

   b) The student observed a spark jumping between one of the charged objects and an earthed conductor.

      Explain this observation. **[2 marks]**

Permanent and induced magnetism, magnetic forces and fields

Poles of a magnet

The poles of a magnet are the places where the magnetic forces are strongest.

When two magnets are brought close together they exert a force on each other.

> **Two like poles repel. Two unlike poles attract.**

Attraction between opposite poles

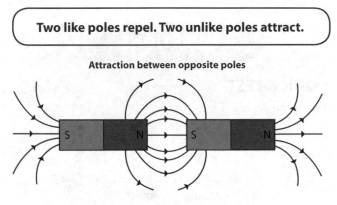

Repulsion between like poles

Neutral or null point

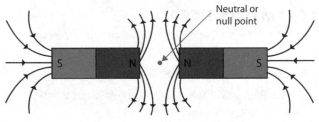

Magnetism is an example of a non-contact force.

Permanent magnetism vs induced magnetism

A permanent magnet . . .

● produces its own magnetic field.

An induced magnet . . .

● becomes a magnet when placed in a magnetic field

● always experiences a force of attraction

● loses most or all of its magnetism quickly when removed from a magnetic field.

Magnetic field

The region around a magnet – where a force acts on another magnet or on a magnetic material (iron, steel, cobalt, magnadur and nickel) – is called the magnetic field.

WS

A compass can be used to plot a magnetic field

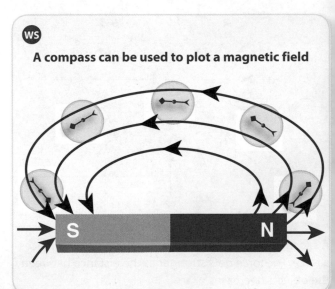

The **force** between a magnet and a magnetic material is always attraction.

The **strength** of the magnetic field depends on the distance from the magnet.

The **field** is strongest at the poles of the magnet.

The **direction** of a magnetic field line is from the north pole of the magnet to the south pole of the magnet.

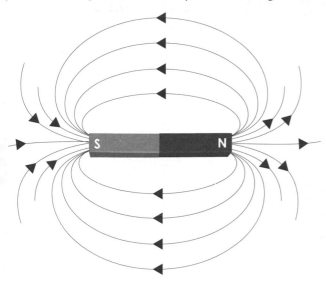

Compasses

A magnetic compass contains a bar magnet that points towards magnetic north. This provides evidence that the Earth's core is magnetic and produces a magnetic field.

SUMMARY

● Opposite poles of a magnet attract each other.

● Like poles of a magnet repel each other.

● A permanent magnet produces its own magnetic field; an induced magnet becomes a magnet when placed in a magnetic field.

● A magnetic field is the area around a magnet.

QUESTIONS

QUICK TEST

1. What happens when the two south poles of a bar magnet are brought together?

2. What happens if opposite poles of two magnets are brought together?

3. What type of force is magnetism?

4. Why does a magnetic compass point north?

EXAM PRACTICE

1. A student was carrying out an investigation into the magnetic field around a bar magnet.

    a) What is a magnetic field? **[1 mark]**

    b) Predict where the magnetic field strength would be the strongest. **[1 mark]**

    c) What force would a magnetic material experience when inside the magnetic field? **[1 mark]**

    d) The student used a compass to plot the magnetic field lines, in what direction would the field lines run? **[1 mark]**

Fleming's left-hand rule, electric motors and loudspeakers

Electromagnets

When a current flows through a conducting wire a magnetic field is produced around the wire.

The shape of the magnetic field can be seen as a series of concentric circles in a plane, perpendicular to the wire.

The direction of these field lines depends on the direction of the current.

The strength of the magnetic field depends on the current through the wire and the distance from the wire.

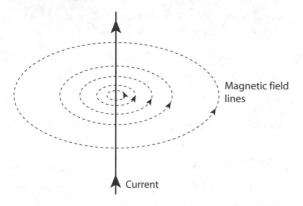

Magnetic field lines

Current

Coiling the wire into a solenoid (a helix) increases the strength of the magnetic field created by a current through the wire.

Magnetic field has a similar shape to that of a bar magnet.

Adding an iron core increases the magnetic field strength of a solenoid.

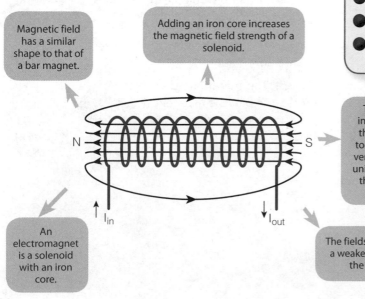

N   S

I_{in} I_{out}

An electromagnet is a solenoid with an iron core.

The fields from individual coils in the solenoid add together to form a very strong, almost uniform, field along the centre of the solenoid.

The fields cancel to give a weaker field outside the solenoid.

🅷🆃 Fleming's left-hand rule and the motor effect

When a conductor carrying a current is placed in a magnetic field, the magnet producing the field and the conductor exert an equal and opposite force on each other. This is the motor effect and is due to interactions between magnetic fields.

The direction of the force on the conductor can be identified using Fleming's left-hand rule.

If the direction of the current or the direction of the magnetic field is reversed, the direction of the force on the conductor is reversed.

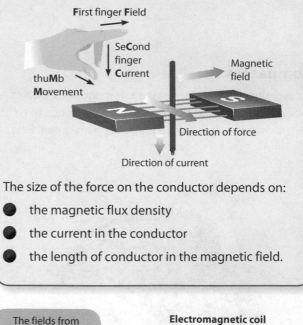

First finger **F**ield

SeCond finger **C**urrent

thu**M**b **M**ovement

Magnetic field

Direction of force

Direction of current

The size of the force on the conductor depends on:

- ● the magnetic flux density
- ● the current in the conductor
- ● the length of conductor in the magnetic field.

Electromagnetic coil

HT For a conductor at right angles to a magnetic field and carrying a current, the force can be calculated using the following equation:

force = magnetic flux density × current × length

$$F = BIl$$

- force, F, in newtons, N
- magnetic flux density, B, in tesla, T
- current, I, in amperes, A (or amp)
- length, l, in metres, m

Example

What is the force produced by a 0.5 m long conductor, with a magnetic flux of 1.2 T and a current of 16 A flowing through it?

$F = BIl$
$F = 1.2 \times 16 \times 0.5 = 9.6$ N

Electric motors

A coil of wire carrying a current in a magnetic field experiences a force, causing it to rotate. This is the basis of an electric motor.

The commutator and graphite brush allow the current to be reversed every half turn to keep the coil spinning.

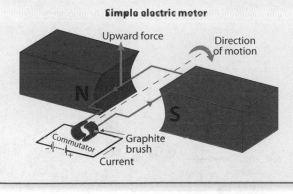

Simple electric motor

Upward force
Direction of motion
N
S
Commutator
Graphite brush
Current

Loudspeakers and headphones

Loudspeakers and headphones use the motor effect to convert variations in current in electrical circuits to the pressure variations in sound waves.

The electrical signal into the coil changes direction. → The electromagnet's magnetic field constantly changes.

↓

This causes the electromagnet to move back and forth, vibrating the cone at different frequencies and generating the sound. ← This leads to changing attraction and repulsion forces between the permanent magnet and the electromagnet.

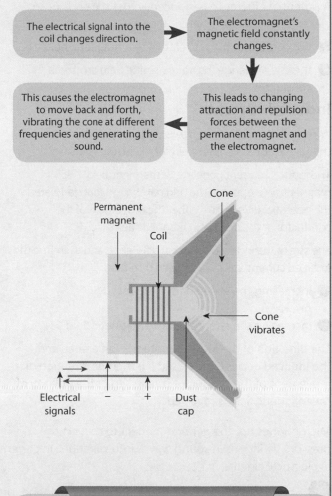

Cone
Permanent magnet
Coil
Cone vibrates
Electrical signals
– +
Dust cap

QUESTIONS

QUICK TEST

1. Name the two things that the strength of a magnetic field around a wire depends on.

HT 2. What three variables are related by Fleming's left-hand rule?

EXAM PRACTICE

HT 1. Calculate the force produced by a 6A current running through a 13 m wire which has a magnetic flux density of 2.5T. **[2 marks]**

SUMMARY

- When a current flows through a conducting wire, a magnetic field is produced around the wire.
- To increase the strength of the magnetic field, the wire can be coiled into a solenoid.
- **HT** An electric motor can be created by placing a wire, carrying a current, in a magnetic field. This will cause the wire to rotate.

Induced potential and transformers

The generator effect

A potential difference is induced across the ends of an electrical conductor when:

- the conductor moves relative to a magnetic field
- there is a change in the magnetic field around a conductor.

Induced current

When a potential difference is induced in a conductor that is part of a complete circuit, a current is induced.

The induced current generates a magnetic field. This magnetic field opposes the original change that generated the potential difference – either the movement of the conductor or the change in magnetic field.

The size of the induced potential difference, and therefore the induced current, can be increased by either:

- increasing the speed of movement
  or
- increasing the strength of the magnetic field.

The direction of the induced potential difference and the induced current is reversed if either the direction of movement of the conductor is reversed or the polarity of the magnetic field is reversed.

Microphones use the generator effect to convert the pressure variations in sound waves into variations in current in electrical circuits.

1 Sound waves move the diaphragm backwards and forwards.

2 The diaphragm moves the coil backwards and forwards.

3 A changing current is induced in the coil that corresponds to the frequency of the sound.

4 This current is used to record the sound or it is fed into an amplifier and into a speaker to produce the sound.

Transformers

A basic transformer has a primary coil wire and a secondary coil of wire wound on an iron core. Iron is used as it is easily magnetised.

WS Uses of the generator effect

The generator effect is used:

- In a dynamo to generate dc. A coil rotates inside the magnetic field of a permanent magnet, inducing a current in the coil. The current always flows in the same direction, producing the dc voltage–time graph below right.

- In an alternator to generate ac. A permanent magnet rotates inside a coil of wire, inducing the current in the coil. The current regularly reverses direction, producing the ac voltage–time graph below left.

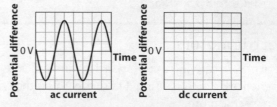

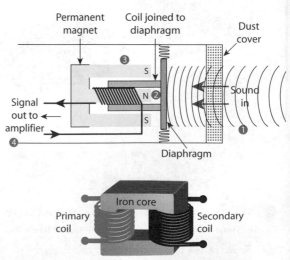

| An alternating current in the primary coil of a transformer produces a changing magnetic field in the iron core. | → | The changing magnetic field is also produced in the secondary coil. | → | An alternating potential difference is induced across the ends of the secondary coil. | → | An induced current will flow in the secondary coil if it is part of a complete circuit. |

The relationship between voltage and number of turns of the coil is given by the following equation:

$$\left[\frac{v_p}{v_s} = \frac{n_p}{n_s}\right]$$

- potential difference on primary and secondary coils, v_p and v_s in volts, V
- number of turns on the primary and secondary coil n_p and n_s

- In a step-up transformer, there are more turns on the secondary coil than the primary coil, so $V_s > V_p$

- In a step-down transformer, there are fewer turns on the secondary coil than the primary coil, so $V_s < V_p$

The power output of a transformer relates to the power input and the potential difference in the coils.

$$V_s \times I_s = V_p \times I_p$$

- V_s and I_s is the power output (secondary coil).
- V_p and I_p is the power input (primary coil).
- power input and output, in watts, W.

Example

The power output required from a transformer is 500 W. The potential difference on the primary coil is 25 V. What current is required to produce this power output?

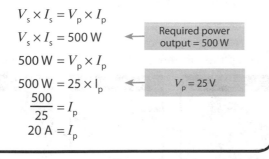

$$V_s \times I_s = V_p \times I_p$$
$$V_s \times I_s = 500 \text{ W} \quad \leftarrow \text{Required power output} = 500 \text{ W}$$
$$500 \text{ W} = V_p \times I_p$$
$$500 \text{ W} = 25 \times I_p \quad \leftarrow V_p = 25 \text{ V}$$
$$\frac{500}{25} = I_p$$
$$20 \text{ A} = I_p$$

If transformers were 100% efficient, the electrical power output would equal electrical power input.

The equation linking potential difference and number of turns can be applied together with the equation for power transfer.

SUMMARY

- **A potential difference is induced across the ends of a conductor when the conductor moves relative to a magnetic field or there is a charge in the magnetic field.**
- **A simple transformer has a primary coil wire and a secondary coil wire wound on an iron core.**

Example

What number of turns on the secondary coil are required to produce a power output of 23 kW, if:

- the potential difference in the primary coil is 400 kV
- the primary coil has 26 000 turns
- the output current is 100 A?

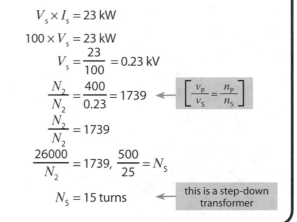

$$V_s \times I_s = 23 \text{ kW}$$
$$100 \times V_s = 23 \text{ kW}$$
$$V_s = \frac{23}{100} = 0.23 \text{ kV}$$
$$\frac{N_2}{N_2} = \frac{400}{0.23} = 1739 \quad \leftarrow \left[\frac{v_p}{v_s} = \frac{n_p}{n_s}\right]$$
$$\frac{N_2}{N_2} = 1739$$
$$\frac{26000}{N_2} = 1739, \quad \frac{500}{25} = N_s$$
$$N_s = 15 \text{ turns} \quad \leftarrow \text{this is a step-down transformer}$$

By transmitting the power at high voltage through power cables, the current is kept low, reducing loss of heat energy. A step-down transformer can then be used to lower the voltage to a safer level.

QUESTIONS

QUICK TEST

1. What is a transformer?

2. Suggest two ways that the induced current from a generator can be increased.

EXAM PRACTICE

1. A transformer had 25 turns on the primary coil and 200 turns on the secondary coil.

   a) Is this an example of a step-up or step-down transformer?

   Explain your answer. **[2 marks]**

   b) If the potential difference across the primary coil is 30 V, what is the potential difference across the secondary coil?

   Show your working. **[3 marks]**

Changes of state and the particle model

The particle model

Matter can exist as a solid, liquid or as a gas.

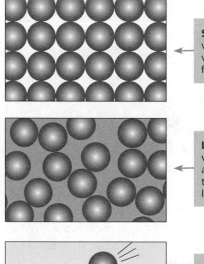

Solid – particles are very close together and vibrating. They are in fixed positions.

Liquid – particles are very close together but are free to move relative to each other. This allows liquids to flow.

Gas – particles in a gas are not close together. The particles move rapidly in all directions.

If the particles in a substance are more closely packed together, the density of the substance is higher. This means that liquids have a higher density than gases. Most solids have a higher density than liquids.

Density also increases when the particles are forced into a smaller volume.

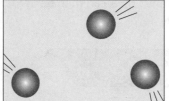

Low density

High density

The density of a material is defined by the following equation:

$$\text{density} = \frac{\text{mass}}{\text{volume}}$$
$$\rho = \frac{m}{v}$$

- density, ρ, in kilograms per metre cubed, kg/m^3
- mass, m, in kilograms, kg
- volume, V, in metres cubed, m^3

Example

What is the density of an object that has a mass of 56 kg and a volume of 0.5 m^3?

$$\rho = \frac{m}{v}$$
$$= \frac{56}{0.5} = 112 \text{ kg/m}^3$$

When substances change state (melt, freeze, boil, evaporate, condense or sublimate), mass is conserved (it stays the same).

Changes of state are physical changes: the change does not produce a new substance, so if the change is reversed the substance recovers its original properties.

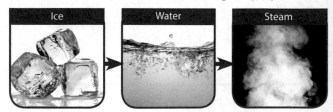

Ice   Water   Steam

Internal energy

Energy is stored inside a system by the particles (atoms and molecules) that make up the system. This is called **internal energy**.

Internal energy of a system is equal to the total kinetic energy and potential energy of all the atoms and molecules that make up the system.

Heating changes the energy stored within the system by increasing the energy of the particles that make up the system. This either raises the temperature of the system or produces a change of state.

Heat and temperature are related but are not a measure of the same thing.

- Heat is the amount of thermal energy and is measured in Joules (J).

- Temperature is how hot or cold something is and is measured in degrees Celsius (°C).

Changes of heat and specific latent heat

When a change of state occurs, the stored internal energy changes, but the temperature remains constant. The graph below shows the change in temperature of water as it is heated; the temperature is constant when the water is changing state.

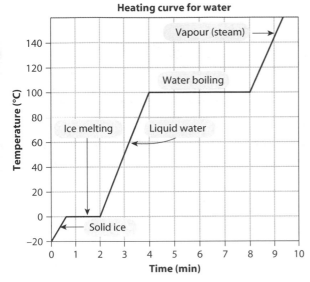

Heating curve for water

The specific latent heat of a substance is equal to the energy required to change the state of one kilogram of the substance with no change in temperature.

The energy required to cause a change of state can be calculated by the following equation:

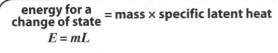

$$\text{energy for a change of state} = \text{mass} \times \text{specific latent heat}$$
$$E = mL$$

- energy, E, in joules , J
- mass, m, in kilograms, kg
- specific latent heat, L, in joules per kilogram, J/kg

Example

What is the energy needed for 600 g of water to melt? (The specific latent heat of water melting is 334 kJ/kg.)

$E = mL$

$0.6 \times 334 = 200.4$ kJ

The specific latent heat of fusion is the energy required for a change of state from solid to liquid.

The specific latent heat of vapourisation is the energy required for a change of state from liquid to vapour.

SUMMARY

- **Matter can exist as solid, liquid or gas.**
- **Specific latent heat of a substance is equal to the energy required to change the state of 1 kg of the substance with no change in temperature.**

QUESTIONS

QUICK TEST

1. What is the density of an object which has a mass of 98 kg and a volume of 0.1 m³?

2. In what three states can matter exist?

3. If a substance is heated but its temperature remains constant, what is happening to the substance?

EXAM PRACTICE

1. a) What energy is needed for 300 kg of iron to turn from a solid to a liquid?

   (The specific latent heat of iron melting is 126 kJ/kg.) **[1 mark]**

   b) What happens to the internal energy and temperature of the iron particles during this process? **[2 marks]**

Particle model and pressure

Gas under pressure

The molecules of a gas are in constant random motion.

When the molecules collide with the wall of their container they exert a force on the wall. The total force exerted by all of the molecules inside the container on a unit area of the wall is the gas pressure.

Increasing the temperature of a gas, held at constant volume, **increases** the pressure exerted by the gas.

Decreasing the temperature of a gas, held at constant volume, **decreases** the pressure exerted by the gas.

The temperature of the gas is related to the average kinetic energy of the molecules. The higher the temperature, the greater the average kinetic energy, and so the faster the average speed of the molecules. At higher temperatures, the particles collide with the walls of the container at a higher speed.

Temperature can be measured in degrees Celsius (°C) or Kelvin (K). To convert from Celsius to Kelvin, add 273.

For example:

● 10°C + 273 = 283 K

0 Kelvin (−273°C) is **absolute zero**. At this point, the particles have no kinetic energy so are not moving.

A gas can be compressed or expanded by pressure changes. The pressure produces a net force at right angles to the wall of the gas container (or any surface).

When the volume of the gas in the container is reduced, the pressure increases.

For a fixed mass of gas held at a constant temperature this equation is used:

> **constant = pressure × volume**
>
> **constant = pV**
>
> ● pressure, p, in pascals, Pa
> ● volume, V, in metres cubed, m^3

When a volume of a gas is altered, this allows the new pressure to be calculated by the following equation:

> $$p_1 \times V_1 = p_2 \times V_2$$
>
> ● p_1 and V_2 are the initial pressure and volume
> ● p_1 and V_2 are the final pressure and volume

Example

A gas with a pressure of 200 kPa is compressed from a volume of 3 m^3 to a volume of 0.5 m^3. What is the new pressure of the gas?

$$p_1 \times V_1 = p_2 \times V_2$$
$$200 \times 3 = p_2 \times 0.5$$
$$600 = p_2 \times 0.5$$
$$\frac{600}{0.5} = p_2$$
$$1200 \text{ kPa} = p_2$$

High pressure – Low volume

Low pressure – High volume

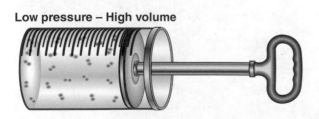

HT Doing work on a gas increases the internal energy of the gas and can cause an increase in the temperature of the gas. Look at this flow chart:

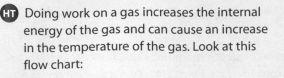

Pumping the handle of a bicycle pump is work done.

↓

Energy is transferred to the air particles in the pump.

↓

This increases the internal energy of the air particles.

↓

As the particles in the air have more internal energy, the temperature of the air in the pump increases.

SUMMARY

● Gas pressure is the force exerted by all the molecules inside a container on a unit area of the wall.

● The temperature of a gas depends on the average kinetic energy of the molecules.

● Temperature can be measured in degrees Celsius or Kelvin. -273°C = 0 Kelvin.

QUESTIONS

QUICK TEST

1. How do gases exert pressure?

2. What is 290 Kelvin in degrees Celsius?

3. What is 28°C in Kelvin?

EXAM PRACTICE

1. A gas syringe was used in an investigation into the effect of volume on gas pressure. The initial volume of the syringe was 0.00001m³ and the initial pressure was 120 kPa.

   a) Why is it important to carry out this investigation at a constant temperature? **[2 marks]**

   b) The gas was compressed by pushing the plunger of the syringe. The new volume of the syringe was 0.000006m³.

      What was the pressure in the syringe now? **[3 marks]**

   c) Pushing the syringe in was an example of doing work on the gas.

      Explain the effect this has on the energy of the air particles in the gas syringe. **[2 marks]**

Atoms and isotopes

Structure of atoms

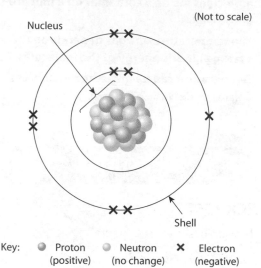

(Not to scale)

Nucleus

Shell

Key: ● Proton (positive) ● Neutron (no change) ✗ Electron (negative)

Atoms have a radius of around 1×10^{-10} metres.

The radius of a nucleus is less than $\dfrac{1}{10\,000}$ of the radius of an atom.

Most of the mass of an atom is concentrated in the nucleus. Protons and neutrons have a relative mass of 1 while electrons have a relative mass of 0.0005. The electrons are arranged at different distances from the nucleus (are at different energy levels).

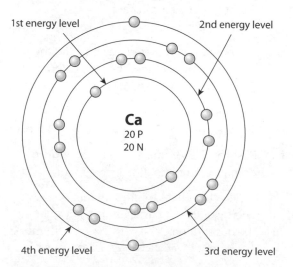

1st energy level

2nd energy level

Ca
20 P
20 N

4th energy level

3rd energy level

Absorption of electromagnetic radiation causes the electrons to become excited and move to a higher energy level and further from the nucleus.

Emission of electromagnetic radiation causes the electrons to move to a lower energy level and move closer to the nucleus.

If an atom loses or gains an electron, it is ionised.

The number of electrons is equal to the number of protons in the nucleus of an atom.

Atoms have no overall electrical charge.

All atoms of a particular element have the same number of protons. The number of protons in an atom of an element is called the **atomic number**.

The total number of protons and neutrons in an atom is called the **mass number**.

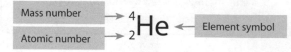

Mass number → $^{4}_{2}$He ← Element symbol
Atomic number →

Atoms of the same element can have different numbers of neutrons; these atoms are called **isotopes** of that element. For example, below are some isotopes of nitrogen. They each have 7 protons in the nucleus but different numbers of neutrons, giving the different isotopes.

$$^{14}N \quad ^{15}N \quad ^{13}N$$

Atoms turn into **positive ions** if they lose one or more outer electrons and into **negative ions** if they gain one or more outer electrons.

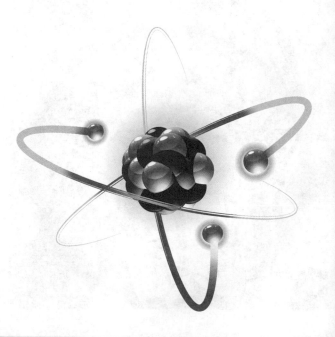

WS The development of the atomic model

Before the discovery of the electron, atoms were thought to be tiny spheres that could not be divided.

The discovery of the electron led to further developments of the model. The plum pudding model suggested that the atom is a ball of positive charge with negative electrons embedded in it.

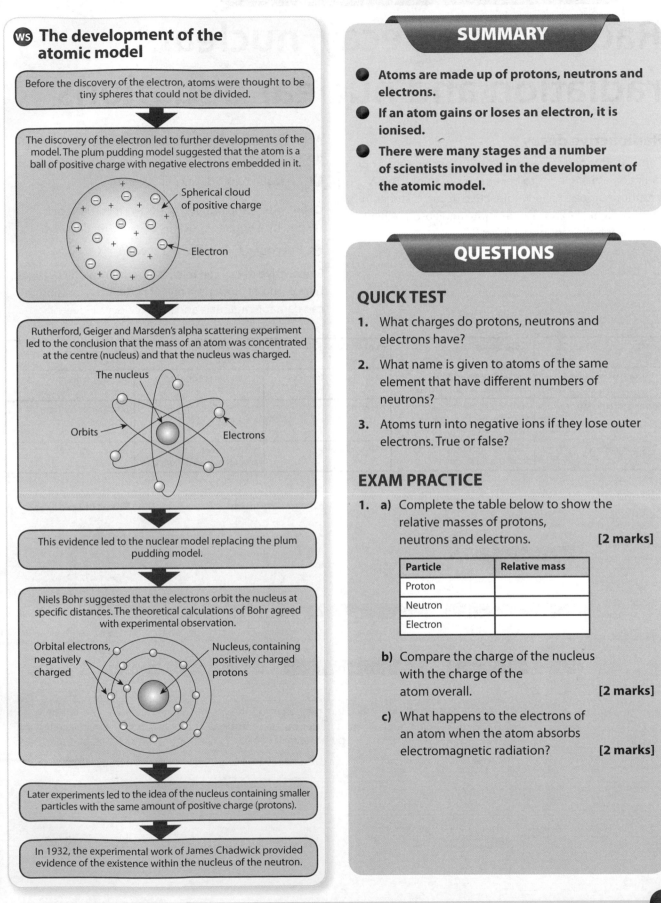

Spherical cloud of positive charge

Electron

Rutherford, Geiger and Marsden's alpha scattering experiment led to the conclusion that the mass of an atom was concentrated at the centre (nucleus) and that the nucleus was charged.

The nucleus

Orbits

Electrons

This evidence led to the nuclear model replacing the plum pudding model.

Niels Bohr suggested that the electrons orbit the nucleus at specific distances. The theoretical calculations of Bohr agreed with experimental observation.

Orbital electrons, negatively charged

Nucleus, containing positively charged protons

Later experiments led to the idea of the nucleus containing smaller particles with the same amount of positive charge (protons).

In 1932, the experimental work of James Chadwick provided evidence of the existence within the nucleus of the neutron.

SUMMARY

- Atoms are made up of protons, neutrons and electrons.
- If an atom gains or loses an electron, it is ionised.
- There were many stages and a number of scientists involved in the development of the atomic model.

QUESTIONS

QUICK TEST

1. What charges do protons, neutrons and electrons have?

2. What name is given to atoms of the same element that have different numbers of neutrons?

3. Atoms turn into negative ions if they lose outer electrons. True or false?

EXAM PRACTICE

1. a) Complete the table below to show the relative masses of protons, neutrons and electrons. **[2 marks]**

Particle	Relative mass
Proton	
Neutron	
Electron	

b) Compare the charge of the nucleus with the charge of the atom overall. **[2 marks]**

c) What happens to the electrons of an atom when the atom absorbs electromagnetic radiation? **[2 marks]**

Radioactive decay, nuclear radiation and nuclear equations

Radioactive decay

Some atomic nuclei are unstable. The nucleus gives out radiation as it changes to become more stable. This is a random process called radioactive decay. Changes in atoms and nuclei can also generate and absorb radiation over the whole frequency range.

Activity is the rate at which a source of unstable nuclei decays, measured in becquerel (Bq).

⚫ 1 becquerel = 1 decay per second

Count rate is the number of decays recorded each second by a detector, such as a Geiger-Müller tube.

⚫ 1 becquerel = 1 count per second

Radioactive decay can release a neutron, alpha particles, beta particles or gamma rays. If radiation is ionising, it can damage materials and living cells.

Particle	Description	Penetration in air	Absorbed by...	Ionising power
Alpha particles (α)	Two neutrons and two protons (a helium nucleus).	a few centimetres	a thin sheet of paper	strongly ionising
Beta particles (β)	High speed electron ejected from the nucleus as a neutron turns into a proton.	a few metres	a sheet of aluminium about 5 mm thick	moderately ionising
Gamma rays (γ)	Electromagnetic radiation from the nucleus.	a large distance	a thick sheet of lead or several metres of concrete	weakly ionising

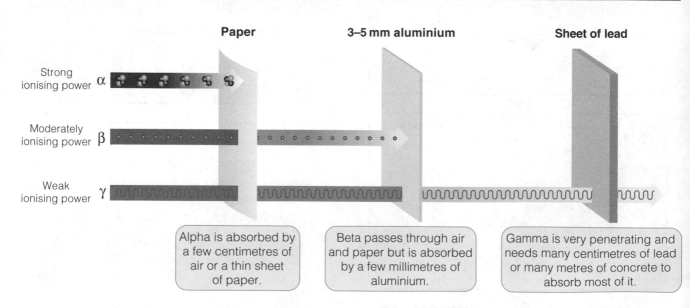

Alpha is absorbed by a few centimetres of air or a thin sheet of paper.

Beta passes through air and paper but is absorbed by a few millimetres of aluminium.

Gamma is very penetrating and needs many centimetres of lead or many metres of concrete to absorb most of it.

Nuclear equations

Nuclear equations are used to represent radioactive decay.

Nuclear equations can use the following symbols:

$$^{4}_{2}\text{He}$$ alpha particle

$$^{0}_{-1}\text{e}$$ beta particle

Alpha decay causes both the mass and charge of the nucleus to decrease, as two protons and two neutrons are released.

$$^{219}_{86}\text{radon} \longrightarrow {}^{215}_{84}\text{polonium} + {}^{4}_{2}\text{He}$$

Beta decay does not cause the mass of the nucleus to change but does cause the charge of the nucleus to change, as a proton becomes a neutron.

$$^{14}_{6}\text{carbon} \longrightarrow {}^{14}_{7}\text{nitrogen} + {}^{0}_{-1}\text{e}$$

The above example is β– decay as a neutron has becomes a proton and an electron has been ejected. In β+ decay a proton becomes a neutron plus a positron.

The emission of a gamma ray does not cause the mass or the charge of the nucleus to change.

Alpha decay

Beta-minus decay with gamma ray

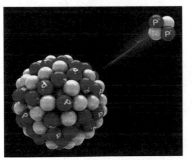

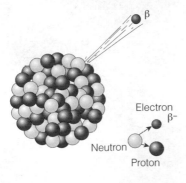

β

Electron
β–

Neutron

Proton

SUMMARY

● **Radioactive decay occurs when unstable atomic nuclei give out radiation as they change to become stable.**
● **The rate at which radioactive decay occurs is measured in Becquerels.**
● **Radioactive decay can release a neutron, alpha particles, beta particles or gamma rays.**
● **Nuclear equations represent radioactive decay.**

QUESTIONS

QUICK TEST

1. How far does beta radiation penetrate in air?

2. What material is required to absorb alpha particles?

3. What effect does beta decay have on the mass and charge of the nucleus of an atom?

EXAM PRACTICE

1. The equation below shows an example of radioactive decay.

$$^{149}_{A}\text{Gd} \rightarrow {}^{B}_{62}\text{Sm} + {}^{4}_{2}\text{He}$$

   a) Identify this type of decay. **[1 mark]**

   b) Give the values of the missing mass numbers A and B. **[2 marks]**

   c) Compare the effect of this type of decay on the mass and charge of the nucleus with the effect of gamma ray emission. **[2 marks]**

Half-lives and the random nature of radioactive decay

Half-life

Radioactive decay occurs randomly. It is not possible to predict which nuclei will decay.

The half-life of a radioactive isotope is the average time it takes for:

● the number of nuclei in a sample of the isotope to halve

or

● the count rate (or activity) from a sample containing the isotope to fall to half of its initial level.

Uranium

Example

A radioactive sample has an activity of 560 counts per second. After 8 days, the activity is 280 counts per second. This gives a half-life of 8 days.

The decay can be plotted on a graph and the half-life determined from the graph.

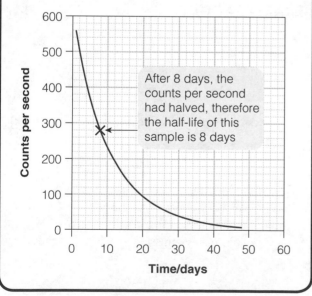

After 8 days, the counts per second had halved, therefore the half-life of this sample is 8 days

HT After another 8 days, the activity would now be 140 counts per second. Therefore, in two half-lives the activity has declined by a total of 560 − 140 = 420 counts per second. As a ratio, the net decline is 420 : 560, or 3 : 4.

Radioactive contamination

Radioactive contamination is the unwanted presence of materials containing radioactive atoms or other materials.

This is a hazard due to the decay of the contaminating atoms. The level of the hazard depends on the type of radiation emitted.

Irradiation is the process of exposing an object to nuclear radiation. This is different from radioactive contamination as the irradiated object does not become radioactive.

Suitable precautions must be taken to protect against any hazard from the radioactive source used in the process of irradiation. In medical testing using radioactive sources, the doses patients receive are limited and medical staff wear protective equipment.

It is important for the findings of studies into the effects of radiation on humans to be published and shared with other scientists. This allows the findings of the studies to be checked by other scientists by the peer review process.

SUMMARY

● Half-life of a radioactive isotope is the average time it takes for the number of nuclei in a sample of the isotope to halve, or the activity from a sample of the isotope to fall to half of its initial level.

● Materials that contain unwanted radioactive atoms or other materials are referred to as radioactive contamination.

● Exposing a material to nuclear radiation is called irradiation.

QUESTIONS

QUICK TEST

1. What is irradiation?

2. Why is radioactive contamination a hazard?

EXAM PRACTICE

1. Researchers carried out an investigation on a radioactive sample which has an activity of 3178 counts per second.

   a) What value will the count rate be after one half life?

   Show how you arrived at your answer. **[2 marks]**

   b) The half life was 6 hours. What will be the count rate after 1 day?

   Show how you arrived at your answer. **[2 marks]**

   c) i) During an investigation into this sample some of the material was found on researchers' clothes.

   Is this an example of contamination or irradiation?

   Explain your answer. **[2 marks]**

   ii) Why is it important that researchers publish the results of their findings on this sample? **[2 marks]**

Hazards and uses of radioactive emissions and background radiation

Background radiation

Background radiation is around us all the time.

Natural radiation sources	Man-made sources
Some rocks, such as granite.	Fallout from nuclear weapons testing.
Cosmic rays from space.	Fallout from nuclear accidents.

The level of background radiation and the radiation dose received by a person may be affected by their occupation and their location.

Radiation dose is measured in sieverts (Sv).

1000 millisieverts (mSv) = 1 sievert (Sv)

Radioactivity can be detected and measured in a number of ways, including using photographic film or a Geiger-Müller tube.

Different half-lives of radioactive isotopes

Radioactive isotopes have a wide range of different half-life values.

Short half-life	Long half-life
Unstable source	Stable source
Rapid decay	Slow decay
Large amount of radiation emitted in a short time	Small amount of radiation emitted per second but continues to emit radiation over a very long period of time
Example: lithium-4 has a half-life of 9.1×10^{23} seconds	Example: calcium-48 has a half-life of 6.4×10^{19} years

Radioactive isotopes with a long half-life can provide a long-term hazard because they continue to emit small amounts of radiation over a long period of time. Radioactive isotopes with a short half-life can provide a short-term danger as large amounts of radiation are released in a short time.

Cosmic Rays

Granite

Uses of nuclear radiation

Radioactivity is used in:

- smoke alarms
- irradiating food to improve its safety and extend shelf life
- sterilisation of equipment, such as medical equipment
- tracing and gauging the thickness of a material, e.g. to monitor the structure of pipes.

Nuclear radiations are used in medicine for:

- exploration of internal organs, such as the use of ^{13}N as a radioactive tracer
- control or destruction of unwanted tissue, e.g. gamma rays are used to destroy cancerous tumours.

A radiation source used as a tracer needs to have a long enough half-life to ensure it emits throughout the test period, but short enough that it doesn't continue to emit after the test period is over.

The radiation emitted should also be able to pass through body tissues so it can be detected but not cause any negative health effects for the patient or medical personnel.

As ionising radiation causes tissue damage, it is important to make sure that it is very well targeted when used to destroy cancer cells. Ionising radiation can also lead to mutations so it's important to balance this risk against the risk posed by the tumour to be destroyed.

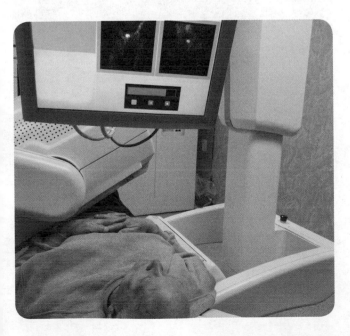

SUMMARY

- **Background radiation comes from natural sources and man-made sources.**
- **Radioactive isotopes with a long half-life can provide a long-term hazard; radioactive isotopes with a short half-life can provide a short-term hazard.**
- **Radioactivity has many uses including in smoke alarms and in sterilising medical equipment.**

QUESTIONS

QUICK TEST

1. Give one man-made source of background radiation.

2. What unit is radiation dose measured in?

3. Give one use of nuclear radiation in medicine.

4. Give one natural source of background radiation.

EXAM PRACTICE

1. Technetium-99m is a radioactive isotope with a half-life of approximately 6 hours, whilst tritium is a radioactive isotope with a half-life of around 12 years.

   Explain which would be the most useful isotope to use as a medical radioactive tracer. **[3 marks]**

Nuclear fission and fusion

Nuclear fission

Nuclear fission is the splitting of a large and unstable atomic nucleus in a radioactive element, such as uranium or plutonium. Spontaneous fission is rare.

Usually, for fission to occur the unstable nucleus must first absorb a neutron.

↓

Some of the mass of the smaller nuclei may be converted into the energy of radiation.

↓

The nucleus emits two or three neutrons plus gamma rays.

↓

Energy is released by the fission reaction.

↓

The neutrons have kinetic energy and may go on to start a chain reaction.

Chain reaction

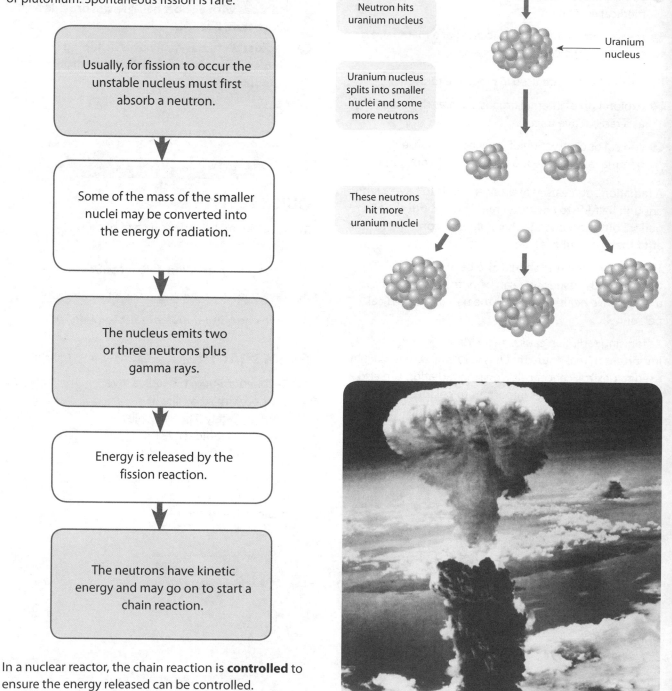

Neutron hits uranium nucleus

Neutron

Uranium nucleus

Uranium nucleus splits into smaller nuclei and some more neutrons

These neutrons hit more uranium nuclei

In a nuclear reactor, the chain reaction is **controlled** to ensure the energy released can be controlled.

In a nuclear weapon, the chain reaction is **uncontrolled**, causing the energy to be released in an explosion.

Nuclear fusion

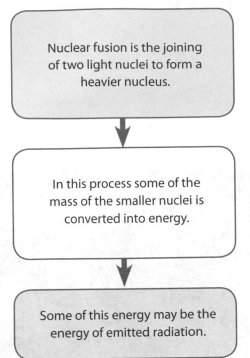

Nuclear fusion is the joining of two light nuclei to form a heavier nucleus.

In this process some of the mass of the smaller nuclei is converted into energy.

Some of this energy may be the energy of emitted radiation.

The protons in the nucleus give it a positive charge. This means that when nuclei are brought close to each other, they repel (electrostatic repulsion).

Very high temperatures and pressures are required to bring the nuclei close enough together and overcome the electrostatic repulsion for fusion to happen. This means that it has not been possible to create a fusion reactor that could be used to generate electricity economically. At the moment, fusion reactors generate less energy than is used to cause the fusion.

Nuclear fusion occurs in stars.

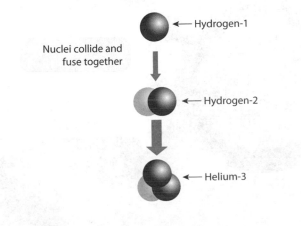

Nuclei collide and fuse together

Hydrogen-1

Hydrogen-2

Helium-3

SUMMARY

- The splitting of large, unstable atomic nuclei is called nuclear fission.
- Nuclear fission can cause a chain reaction to occur.
- The joining of two light nuclei to form a heavier nucleus is called nuclear fusion.
- Nuclear fusion occurs in stars.

QUESTIONS

QUICK TEST

1. What is the difference between the release of energy in a nuclear reactor and from a nuclear weapon?

2. Explain why very high temperatures and pressures are required to cause fusion to happen.

3. In nuclear fusion, heavier nuclei join to form lighter nuclei. True or false?

EXAM PRACTICE

1. The equation below shows an example of a nuclear reaction:

$$^{12}_{6}C + ^{1}_{1}H \rightarrow ^{13}_{7}N$$

   a) Identify the type of reaction and explain your answer. **[2 marks]**

   b) Is this type of reaction used to generate electricity at the moment?

   Explain your answer. **[2 marks]**

Our solar system and the life cycle of a star

The planets in the solar system orbit the Sun (a star). Dwarf planets, such as Pluto, which also orbit the Sun and the natural satellites of planets (moons), are also part of the solar system.

NEPTUNE

URANUS

SATURN

JUPITER

MARS

EARTH

VENUS

MERCURY

Not to scale

The Sun

The Sun was formed from dust and gas (nebula) pulled together by gravitational attraction. Collisions between particles caused the temperature to increase enough for hydrogen nuclei to fuse together forming helium.

The energy released by nuclear fusion processes keeps the core of the Sun hot.

The Sun is stable. The force of gravity acting inwards and trying to collapse the Sun is in equilibrium (balanced) with the outward force produced by the fusion energy trying to expand the Sun.

The Sun is in the main sequence period of its life cycle.

The life cycle of a star

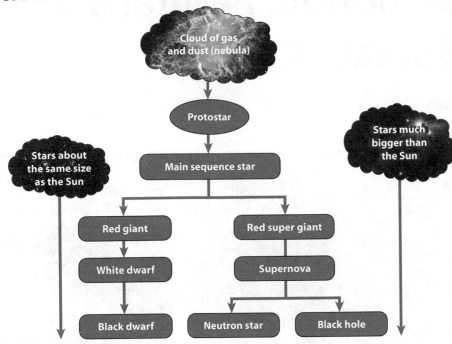

- Fusion processes in stars produce all of the naturally occurring elements.
- All elements heavier than iron are produced in a supernova.
- The explosion of a supernova distributes the elements throughout the Universe.

SUMMARY

- **The planets in our solar system orbit the Sun.**
- **The Sun was formed from dust and nebula pulled together by gravitational attraction.**
- **Nuclear fusion takes place in the Sun and all other stars.**
- **Stars go through a life cycle, ending as a black dwarf, a neutron star or a black hole.**

QUESTIONS

QUICK TEST

1. What stage in its life cycle is the Sun currently in?

2. What three things can a star end up as at the end of its life cycle?

3. Would a star bigger than the Sun form a supernova or a white dwarf?

QUESTIONS

EXAM PRACTICE

1. a) Compare how the elements magnesium and iridium are formed.

   Use the periodic table to help with your answer. **[2 marks]**

   b) Magnesium and iridium are found throughout the galaxy.

   Explain how they were distributed in this way. **[1 mark]**

2. Explain the balance of forces that the Sun is currently experiencing. **[2 marks]**

Orbital motion, natural and artificial satellites

Natural satellites

Planets orbit the Sun and a moon orbits a planet.

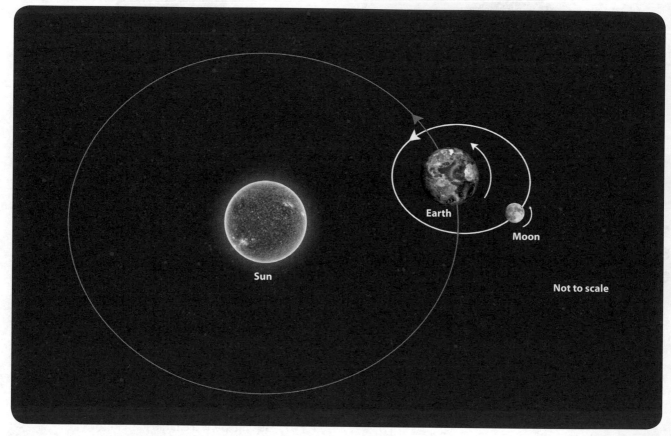

Sun

Earth

Moon

Not to scale

Man-made satellites

Artificial (man-made) satellites orbit the Earth and have a variety of functions including communication and weather monitoring.

Satellites in geostationary orbits always stay above the same point of the Earth. Satellites in polar orbits travel over both poles during an orbit. These orbits often only take a few hours.

Gravity

Provides the force that allows planets and satellites to maintain their circular orbits.

HT

- The force of gravity acts towards the centre of the circular orbit.
- This causes the object to accelerate in that direction.
- The instantaneous velocity of the orbiting body is at a right angle to the direction of the force of gravity.
- The velocity of the orbiting body is constantly changing but the speed is constant. If the body changes speed the orbital height will change.

Red-shift

Our solar system is a small part of the Milky Way galaxy. There are hundreds of billions of other galaxies in the Universe.

There is an increase in the wavelength of light from most distant galaxies.
As the wavelength of visible light increases this leads to the light shifting to the red end of the spectrum. This effect is called red-shift.

↓

Galaxies are moving away from the Earth. The further away the galaxies, the faster they are moving. Observations of supernovae suggest that distant galaxies are receding ever faster.

↓

The faster movement leads to a larger observed increase in wavelength and so a larger red-shift.

WS This observed red-shift provides evidence that the Universe is expanding and also supports the Big Bang theory. The Big Bang theory states that the Universe began from a very small point that was extremely hot and dense. Cosmic background radiation (CMB) also provides evidence for the Big Bang theory.

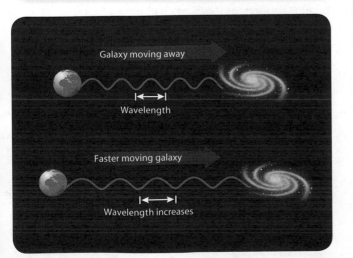

Galaxy moving away
Wavelength

Faster moving galaxy
Wavelength increases

SUMMARY

- Planets and moons are natural satellites.
- Planets orbit the Sun and a moon orbits a planet.
- Man-made satellites orbit the Earth. Man-made satellites are used for communication and weather monitoring.
- Gravity is a force that allows satellites to maintain their orbit.
- Red-shift provides evidence that the Universe is expanding, and supports the Big Bang theory.

QUESTIONS

QUICK TEST

1. What is gravity?
2. Give two uses of artificial satellites.
3. What happens to the wavelength of light from a galaxy that is moving away from the Earth?
4. What galaxy is our solar system part of?
5. What is CMB?

EXAM PRACTICE

1. The static universe theory is a cosmological model in which the universe is infinite in size and has existed and will exist for an infinite period of time.

   A supporter of this theory was the 16th Century astronomer, Thomas Digges.

   Explain why this theory is no longer considered to be correct.

   Include two pieces of evidence in your answer. **[4 marks]**

Answers

Page 5

QUICK TEST

1. A vector quantity has both a magnitude and a direction.
2. A scalar quantity only has a magnitude.
3. $0.033 \times 10 = 0.33$ N
4. 57 J

EXAM PRACTICE

1. a) Mass = weight / gravitational field strength
   $$= 560 / 9.8 \text{ [1]}$$
   $$= 57.1 \text{kg [1]}$$
   b) Work done = force × distance moved
   $$= 560 \times 4 \text{ [1]}$$
   $$= 2240 \text{J [1]}$$
   c) Pushing the crate is a contact force as two objects are physically touching [1] whilst magnetic force does not require objects to be physically touching. [1]

Page 7

QUICK TEST

1. A stretched object that returns to its original length after the force is removed
2. $0.3 \times 2 = 0.6$ N
3. Elastic potential energy

EXAM PRACTICE

1. a) Extension = force / spring constant
   $$= 7 / 18 \text{ [1]}$$
   $$= 0.39 \text{m [1]}$$
   b) i) A straight line [1]
      ii) A linear relationship [1]
      iii) A non-linear relationship [1] as all the points could no longer be connected by a straight line [1]

Page 9

QUICK TEST

1. True
2. The turning effect of a force
3. $50 \times 0.4 = 20$ Nm

EXAM PRACTICE

1. a) Distance to pivot = moment of force / force
   $$= 600 / 400 \text{ [1]}$$
   $$= 1.5 \text{ Nm [1]}$$
   b) Moment of force to balance seesaw = 600 Nm [1]
   Distance to pivot = moment of force / force
   $$= 600 / 300$$
   $$= 2 \text{ m [1]}$$
2. The speed of rotation would be slower than 60 rotations per second [1] as in a gear system a larger gear will rotate slower than the smaller gear it is interlocking with. [1]

Page 11

QUICK TEST

1. As the height of an object increases there are fewer air molecules above the object so their total weight is smaller.
2. $19 \times 1000 \times 10 = 190\,000$ Pa = 190 kPa
3. The upward force that a liquid or gas exerts on a body floating in it.

EXAM PRACTICE

1. a) Liquids are relatively incompressible [1] whilst gases are compressible. [1]
   b) Force = pressure × area of that force
   $$= 1200 \times 0.2 \text{ [1]}$$
   $$= 240 \text{N [1]}$$
2. a) The force of the boat's weight was greater [1] than the upthrust so the boat sank. [1]
   b) Reduce the density of the boat [1] so it was less dense than the liquid it was being floated on. [1]

Page 13

QUICK TEST

1. Speed has a magnitude but not a direction.
2. $\frac{15}{3} = 5$ km/h
3. 330 m/s

EXAM PRACTICE

1. a) Distance = speed × time
   = 21 × 56 **[1]**
   = 1176m **[1]**

   b) Time = distance / speed
   = 14000 / 15 **[1]**
   = 933 seconds **[1]**
   933 / 60 = 15.6 minutes
   Time = 15.6 minutes **[1]**

   c) Her velocity changes **[1]** as although her speed remains constant the direction she is travelling in is changing. **[1]**

Page 15

QUICK TEST

1. Change in velocity over time
2. A negative acceleration/deceleration
3. Speed

EXAM PRACTICE

1. a) Change in velocity = 54-0
   = 54m/s **[1]**
   Average acceleration = 54 / 16 = 3.4m/s² **[1]**

   b) They have reached terminal velocity **[1]** so the resultant force is 0. **[1]**

Page 17

QUICK TEST

1. 93 × 10 = 930 N
2. 83 N
3. It continues to move at the same speed.
4. It remains stationary.

EXAM PRACTICE

1. a) Acceleration = force / mass
   = 2100 / 7 **[1]**
   = 300 m/s² **[1]**

   b) 2100N **[1]** as Newton's third law states **[1]** that when two objects interact they exert equal and opposite forces on each other. **[1]**

Page 19

QUICK TEST

1. 0.2–0.9 seconds
2. Thinking distance and braking distance
3. Tiredness, drugs, alcohol, distractions (e.g. eating, drinking, using a mobile phone).

EXAM PRACTICE

1. a) Thinking distance **[1]**

   b) No **[1] Two from**: Other factors affect stopping distance **[1]** including speed of the vehicle **[1]** adverse road conditions or vehicle condition. **[1]**

2. a) Work is done by the friction force between the brakes and the wheel **[1]** this reduces the kinetic energy of the vehice **[1]** until it reaches 0 and the vehicle comes to a stop. **[1]**

   b) Very heavy braking leads to a large rise in the temperature of the brakes due to friction, causing them to smoke. **[1]**

Answers

Page 21
QUICK TEST

1. Seat belts; Airbags; **any other suitable answer**

2. Airbags slow down a person's change in momentum in a collision.

3. $8 \times 5 = 40$ kg m/s

EXAM PRACTICE

1. Initial momentum = mass x velocity
    $$= 75 \times 9$$
    $$= 675 \text{ kg m/s } [1]$$

    Final momentum = 0

    Change in momentum = 675 - 0 = 675 kg m/s [1]

    Force = change in momentum / time
    $$= 675 / 0.1$$
    $$= 6750 \text{ N } [1]$$

    b) i) The force would be higher [1] as the driver would have come to rest in a much faster time if he was not wearing a seat belt, increasing the rate of change in momentum. [1]

    ii) The force would be lower [1] as the driver would have come to rest in a slower time, slowing down the change in momentum. [1]

Page 23
QUICK TEST

1. It is the amount of energy required to raise the temperature of one kilogram of a substance by one degree Celsius.

2. $0.5 \times 0.62 \times 5^2 = 7.75$ J

3. $17 \times 456 \times 10 = 77.52$ kJ

EXAM PRACTICE

1. a) Elastic potential energy = 0.5 × spring constant × extension^2
    $$= 0.5 \times 115 \times 0.49^2 \text{ [1]}$$
    $$= 13.8\text{N } [1]$$

    b) Elastic potential to kinetic energy [1] and gravitational potential energy. [1]

    c) 0 [1] as the ball is no longer raised above the ground. [1]

Page 25
QUICK TEST

1. Material used to reduce transfer of heat energy

2. Power = work done/time
    $$= 50/2$$
    $$= 25 \text{ W}$$

3. It reduces the thermal conductivity.

EXAM PRACTICE

1. $\frac{600}{802} = 0.75 = 75\%$ [1]

Page 27
QUICK TEST

1. **Three from**: Biofuel; Wind; Hydro-electricity; Geothermal; The tides (tidal power); The sun (solar power); Water waves

2. **Three from**: Coal; Oil; Gas; Nuclear fuel

3. The wind doesn't always blow and it's not always sunny, so electricity isn't always generated.

4. They produce carbon dioxide, which is a greenhouse gas. Increased greenhouse gas emissions are leading to climate change. Particulates and other pollutants are also released, which cause respiratory problems.

EXAM PRACTICE

1. a) Renewable energy resources will not run out [1] and do not produce carbon dioxide, which is a greenhouse gas that can lead to climate change. [1] Nuclear power plants also do not produce carbon dioxide [1] and are very reliable so can generate electricity even if renewable energy resources are not, e.g. if the wind isn't blowing. [1]

    b) Nuclear power plants produce hazardous nuclear waste [1] and have the potential for devastating nuclear accidents. [1]

    c) Some people consider wind farms to be ugly and to spoil the landscape [1] building tidal power stations can destroy important tidal habitats. [1]

Page 29

QUICK TEST

1. The distance from a point on one wave to the equivalent point on an adjacent wave
2. **One from:** Water wave; Electromagnetic wave; S-wave
3. **One from:** Sound wave; P-wave
4. Amplitude, wavelength, frequency, period

EXAM PRACTICE

1. **a)** Water waves **[1]**

   b) period = 1 / frequency

   $$= 1 / 2 \text{ [1]}$$

   $$= 0.5 \text{ seconds [1]}$$

   c) wave speed = frequency × wavelength

   $$= 2 \times 0.3 \text{ [1]}$$

   $$= 0.6 \text{ m/s [1]}$$

   d) This shows that waves transfer energy and information **[1]** without transferring matter. **[1]**

Page 31

QUICK TEST

1. The angle between the incident ray and the normal
2. Ultrasound
3. The angle of reflection
4. It could be absorbed, transmitted or reflected.

EXAM PRACTICE

1. **a)** Sound waves cause the eardrum to vibrate. **[1]** If it is torn it will be unable to vibrate properly; **[1]** this will mean the vibrations will not be properly passed to the brain to be perceived as sounds. **[1]**

   b) As people age their hearing range shrinks **[1]** with a reduction in the upper range **[1]** 18KHz is in the upper range of human hearing. **[1]**

Page 33

QUICK TEST

1. P-waves are longitudinal and travel at different speeds through solids and liquids. S-waves are transverse and cannot travel through a liquid.
2. Longitudinal waves with a frequency higher than the upper limit of hearing for humans.
3. Sound wave

EXAM PRACTICE

1. P-waves are longitudinal, seismic waves which can travel through solids and liquids. **[1]** S-waves are transverse, seismic waves which can travel through solids but not liquids. **[1]** The outer core of the Earth's surface is liquid. **[1]** This means the P-waves pass through and are detected on the opposite side of the Earth whilst S-waves are not. **[1]**

2. Distance = speed × time

   $$= 1500 \times 1.3$$

   $$= 1950\text{m [1]}$$

   1950 / 2 = 975m **[1]**

Page 35

QUICK TEST

1. Infrared
2. Visible light
3. Transverse
4. The change in direction of a wave as it travels from one medium to another
5. Towards the normal

EXAM PRACTICE

1. Visible light and UV radiation are both examples of electromagnetic waves. **[1]** All electromagnetic waves travel at the same speed through the vacuum of space. **[1]**

2. Light refracts as it moves between the water and the air. **[1]** As light travels faster in air than water the light bends away from the normal as it moves from water to the air. **[1]** This means the fish are not exactly where they appear to be. **[1]**

Answers

Page 37
QUICK TEST
1. 1000 millisieverts = 1 sievert
2. By oscillations in electrical circuits
3. Sterilising, medical imaging, treating cancer
4. **One from**: Electrical heaters; Cooking food; Infrared cameras; Television remote controls

EXAM PRACTICE
1. X-rays are absorbed differently by different parts of the body. **[1]** More are absorbed by hard tissues e.g. bones **[1]** and less are absorbed by soft tissues. This allows an image of the inside of the body to be created. **[1]**
2. Ultraviolet radiation can lead to premature ageing of skin **[1]** and increased risk of skin cancer. **[1]**

Page 39
QUICK TEST
1. An object that reflects light or absorbs all light that falls on it.
2. A virtual image
3. Object height = image height/magnification
$$= 22/8$$
$$= 2.75 \text{ cm}$$

EXAM PRACTICE
1. a) The filters are transparent **[1]** as they don't diffuse the light. **[1]**
   b) The blue object would appear black. **[1]** As none of the blue light reflected from it would pass the through the red filter. **[1]**
   c) Diffuse reflection occurs when light is scattered from a rough surface. **[1]** Specular reflection occurs when light is reflected from a smooth surface in a single direction. **[1]**

Page 41
QUICK TEST
1. It absorbs all of the radiation incident on it, it does not reflect or transmit any radiation and it is the best possible emitter as it emits the maximum amount of radiation possible at a given temperature.
2. No, it's impossible for a body in the real world to fulfill the criteria, for a perfect black body
3. Its temperature will remain constant.

EXAM PRACTICE
1. a) i) The metal was absorbing more radiation than it was emitting. **[1]**
      ii) The metal was emitting more radiation than it was absorbing. **[1]**
      iii) The radiation emitted by the metal was equal to the radiation being absorbed by the metal. **[1]**
   b) When its temperature was very high. **[1]**

Page 43
QUICK TEST
1. —▭—
2. 3.8 C
3. 600/3 = 200 A

EXAM PRACTICE
1. a) A battery or a cell **[1]** as this is a source of energy which produces a potential difference. **[1]**
   b) i) Time = charge flow / current
      $$= 160 / 8 \text{ [1]}$$
      $$= 20 \text{ seconds [1]}$$
      ii) 8A **[1]** as the current is the same at any point in a closed circuit. **[1]**

Page 45
QUICK TEST
1. Resistance depends on light intensity so current is not directly proportional to potential difference.
2. $6 \times 3 = 18$ V
3. 8/2 = 4

EXAM PRACTICE
1. a) An ammeter **[1]** and a voltmeter **[1]**
   b) Current = potential difference / resistance
      $$= 12 / 4 \text{ [1]}$$
      $$= 3A \text{ [1]}$$
   c) The temperature of the filament increased. **[1]** This would cause the resistance of the lamp to increase. **[1]**

Page 47
QUICK TEST
1. Add the resistances of both resistors together.
2. 4 A
3. A circuit that contains branches and where all the components have the same potential difference across them.

EXAM PRACTICE
1. a) 9V [1] as all the components in a parallel circuit have the same potential difference across them. [1]
   b) The total resistance of all the components is less than 3 ohms [1] because in a parallel circuit the total resistance of all the components is less than the resistance of the smallest individual resistor. [1]

Page 49
QUICK TEST
1. Alternating current and direct current
2. Circuit breakers can be reset and operate faster than a fuse.
3. Brown
4. 0V

EXAM PRACTICE
1. a) No [1] The appliance was double insulated [1] which means it is impossible for the case to become live as it's plastic. [1] The appliance therefore did not require an earth wire. [1]
   b) In the case of a fault a large current flows from the live wire to the earth wire. [1] This melts the fuse and disconnects the live wire. [1]

Page 51
QUICK TEST
1. They lower the potential difference of the transmission cables to a safe level for domestic use.
2. $50 \times 10 = 500$ J
3. $4^2 \times 3 = 48$ W

EXAM PRACTICE
1. a) Resistance = power / current^2
   $= 8000000 / 500^2$ [1]
   $= 32$ ohms [1]
   b) Energy transferred = power × time
   $= 8000000 \times 120$ [1]
   $= 960000$ kJ [1]
   c) The current is kept relatively low to reduce energy loss due to heating in the cables. [1] This increases the efficiency of the energy transmission. [1]

Page 53
QUICK TEST
1. An imbalance of electrical charges on/in a material
2. The loss of electrons
3. They will repel

EXAM PRACTICE
1. a) Both objects would have electrical fields. [1] These are areas where a charged object experiences a force [1] as one object was positive and the other was negative; when they entered each other's electrical fields they would both experience a force of attraction. [1]
   b) There was a large potential difference [1] between the charged object and the earthed conductor, which caused the spark. [1]

MAGNETISM AND ELECTROMAGNETISM
Page 55
QUICK TEST
1. They repel
2. They attract
3. A non-contact force
4. The Earth's core is magnetic and produces a magnetic field.

EXAM PRACTICE
1. a) The region around a magnet where a force acts on another magnet or on magnetic material. [1]
   b) At the poles of the magnet [1]
   c) A force of attraction [1]
   d) From the north pole to the south pole [1]

Page 57
QUICK TEST
1. The current through the wire and the distance from the wire
2. Magnetic field, force and current

EXAM PRACTICE
1. Force = magnetic flux density × current × length
   $= 2.5 \times 6 \times 13$ [1]
   $= 195$N [1]

Answers

QUICK TEST
1. A device used to increase/decrease the voltage of an alternating current
2. By increasing the speed of movement or increasing the strength of the magnetic field

EXAM PRACTICE
1. a) This is a step-up transformer **[1]** as it has more turns on the secondary coil than the primary coil. **[1]**

   b) $n_p / n_s = V_p / V_s$

   $25 / 200 = 30 / V_s$

   $0.125 = 30 / V_s$ **[1]**

   $V_s = 30 / 0.125$ **[1]**

   $V_s = 240V$ **[1]**

Page 61
QUICK TEST
1. $98/0.1 = 980$ kg/m^3
2. Solid, liquid or gas
3. It is changing state

EXAM PRACTICE
1. a) $300 \times 126 = 37\ 800$ kJ **[1]**
   b) The internal energy of the particles increases **[1]** whilst the temperature remains constant. **[1]**

Page 63
QUICK TEST
1. When the molecules collide with the wall of their container they exert a force on the wall, causing pressure.
2. 17°C
3. 301K

EXAM PRACTICE
1. a) Temperature has an effect on pressure. **[1]** Temperature must be kept constant to show that it is the volume that is effecting the pressure and not the temperature. **[1]**

   b) $P_1 \times V_1 = P_2 \times V_2$

   $120 \times 0.00001 = P_2 \times 0.000006$ **[1]**

   $0.0012 = P_2 \times 0.000006$ **[1]**

   $P_2 = 0.0012 / 0.000006$

   $P_2 = 200$ Kpa **[1]**

   c) Energy is transferred to the air particles. **[1]** This increases their internal energy. **[1]**

Page 65
QUICK TEST
1. Protons – positive; Neutrons – no charge; Electrons – negative
2. Isotopes
3. False. Atoms turn into positive ions if they lose outer electrons.

EXAM PRACTICE
1. a)

Particle	Relative mass
Proton	1
Neutron	1
Electron	0.0005

   [All correct 2 marks, 1 mark if 1 incorrect]

   b) The nucleus is positively charged **[1]** whilst the atom has no overall charge. **[1]**

   c) They become excited and move to a higher energy level **[1]** further from the nucleus. **[1]**

Page 67
QUICK TEST
1. A few metres
2. A thin sheet of paper
3. The mass of the nucleus doesn't change but the charge of the nucleus does change.

EXAM PRACTICE
1. a) Alpha decay **[1]**

   b) A = 64 **[1]** B = 145 **[1]**

   c) Alpha decay causes the mass and charge of the nucleus to decrease. **[1]** Gamma ray emission does not cause the mass or charge of the nucleus to change. **[1]**

Page 69
QUICK TEST
1. Exposing an object to nuclear radiation without the object becoming radioactive itself.
2. The radioactive atoms decay and release radiation.

EXAM PRACTICE
1. a) $3178 / 2 = 1589$ **[2]**
    b) $3178 / 2 = 1589$ = count after 6 hours
       $1589 / 2 = 794.5$ **[1]** count after 12 hours
       $794.5 / 2 = 397.25$ count after 18 hours
       $397.25 = 198.625$ **[1]** count after 24 hours
    c) i) This is contamination **[1]** as it's the unwanted presence of materials containing radioactive atoms. **[1]**
       ii) So the results can be checked by other scientists **[1]** in the peer review process **[1]**

Page 71
QUICK TEST
1. **One from**: Fallout from nuclear weapons testing; Fallout from nuclear accidents; X-rays
2. Sieverts
3. **One from**: Killing cells (such as cancer); Exploring internal organs (such as tracers)
4. **One from**: Some rocks (such as granite); Cosmic rays from space

EXAM PRACTICE
1. Technetium-99m **[1]** as it has the shortest half-life **[1]** as it will emit radiation during the test period but not for a long time afterwards. **[1]**

Page 73
QUICK TEST
1. The energy release from a nuclear reactor is controlled. The energy release from a nuclear weapon is uncontrolled.
2. To overcome the electrostatic repulsion between the protons in the nucleus
3. False. Lighter nuclei join to form heavier nuclei.

EXAM PRACTICE
1. a) Nuclear fusion **[1]** Two nuclei are joining to form a heavier nuclei. **[1]**
    b) No **[1]** because at the moment fusion reactors generate less energy than is used to start the fusion reaction. **[1]**

Page 75
QUICK TEST
1. Main sequence
2. Black dwarf, neutron star or black hole
3. Supernova

EXAM PRACTICE
1. a) As its lighter than iron, magnesium is formed by fusion in a star. **[1]** Iridium is heavier than iron so is formed by the supernova of a star. **[1]**
    b) Supernovas of stars have distributed these elements throughout the galaxy. **[1]**
2. The expansion forces produced by the fusion reactions in the Sun **[1]** are balanced by the gravity acting inwards trying to collapse the Sun. **[1]**

Page 77
QUICK TEST
1. The process where all objects with mass attract each other
2. **Two from**: Weather monitoring; Spying; Communications
3. The wavelength increases and is shifted to the red end of the spectrum.
4. Milky Way
5. Cosmic background radiation

EXAM PRACTICE
1. The Big Bang theory states the universe started from a very small point that was extremely hot and dense. **[1]** The universe has been expanding since the big bang. **[1]** The evidence for this is the red-shift of galaxies moving away from the Earth **[1]** and cosmic background radiation. **[1]**

Glossary

ac – Alternating current that changes direction

Acceleration – Change in velocity over time

Atomic number – Number of protons in an atom

Background radiation – Radiation which is around us all the time

Becquerel – Unit of rate of radioactive decay

Braking distance – Distance a vehicle travels after the brakes have been applied

Compression – A region in a longitudinal wave where the particles are closer together

dc – Direct current that always passes in the same direction

Displacement – Distance travelled in a given direction

Elastically deformed – Stretched object that returns to its original length after the force is removed

Electric current – Flow of electrical charge

Emission – Release of energy

Gas pressure – Total force exerted by all of the gas molecules inside the container on a unit area of the wall

Gravity – Process where all objects with mass attract each other

Inelastically deformed – Stretched object that does not return to its original length after the force is removed

HT Inertia – The tendency of objects to continue in their state of rest or uniform motion

Irradiation – Exposing an object to nuclear radiation without the object becoming radioactive itself

Magnetic field – Region around a magnet where a force acts on another magnet or on a magnetic material

Mass number – Total number of protons and neutrons in an atom

Moment – Turning effect of a force

Momentum – Product of an object's mass and velocity

National Grid – System of transformers and cables linking power stations to consumers

Non-renewable – Energy resource that will eventually run out

Nuclear fission – Splitting of a large and unstable atomic nucleus in a radioactive element

Nuclear fusion – Joining of two light nuclei to form a heavier nucleus

Ohmic conductor – Resistor at constant temperature where the current is directly proportional to the potential difference

Opaque – Object that reflects light

Parallel circuit – Circuit which contains branches and where all the components will have the same voltage

Permanent magnet – Magnet which produces its own magnetic field

Radioactive decay – Random release of radiation from an unstable nucleus as it becomes more stable

Rarefaction – A region in a longitudinal wave where the particles are further apart

Reflection – Waves striking a boundary between different media and being returned back to the same media

Refraction – Change in direction of a wave as it travels from one medium to another

Renewable – Energy resource that can be replenished as it is used

Satellite – Object in orbit around a body

Scalar – A quantity that only has a magnitude

Series circuit – Circuit where all components are connected along a single path

Sievert – Unit of radiation dose

Solar system – Bodies orbiting a star

Solenoid – Coil wound into a helix shape

Specific heat capacity – Amount of energy required to raise the temperature of one kilogram of a substance by one degree Celsius

Specific latent heat – The energy required to change the state of one kilogram of the substance with no change in temperature

Star – Very large body of gas held together by its own gravity

Static electricity – Imbalance of electrical charges on or in a material

Thermal insulation – Material used to reduce transfer of heat energy

Thinking distance – Distance travelled before the driver applies the brakes

HT **Transformer** – Device used to increase or lower the voltage of an alternating current

Translucent – Object that diffuses the light that passes through it, meaning that things observed

Transparent – Object that allows all light to pass through

Ultrasound – Longitudinal waves with a frequency higher than the upper limit of hearing for humans

HT **Upthrust** – Upward force that a liquid or gas exerts on a body in a fluid floating in it

Vector – A quantity that has both a magnitude and a direction

Velocity – Speed in a given direction

Weight – Force acting on an object due to gravity

The Periodic Table

Key
- Metals
- Non-metals

Key box:
- Relative atomic mass → 1
- Atomic symbol → H
- Name → hydrogen
- Atomic/proton number → 1

1	2											3	4	5	6	7	0 or 8
																	4 **He** helium 2
7 **Li** lithium 3	9 **Be** beryllium 4											11 **B** boron 5	12 **C** carbon 6	14 **N** nitrogen 7	16 **O** oxygen 8	19 **F** fluorine 9	20 **Ne** neon 10
23 **Na** sodium 11	24 **Mg** magnesium 12											27 **Al** aluminium 13	28 **Si** silicon 14	31 **P** phosphorus 15	32 **S** sulfur 16	35.5 **Cl** chlorine 17	40 **Ar** argon 18
39 **K** potassium 19	40 **Ca** calcium 20	45 **Sc** scandium 21	48 **Ti** titanium 22	51 **V** vanadium 23	52 **Cr** chromium 24	55 **Mn** manganese 25	56 **Fe** iron 26	59 **Co** cobalt 27	59 **Ni** nickel 28	63.5 **Cu** copper 29	65 **Zn** zinc 30	70 **Ga** gallium 31	73 **Ge** germanium 32	75 **As** arsenic 33	79 **Se** selenium 34	80 **Br** bromine 35	84 **Kr** krypton 36
85 **Rb** rubidium 37	88 **Sr** strontium 38	89 **Y** yttrium 39	91 **Zr** zirconium 40	93 **Nb** niobium 41	96 **Mo** molybdenum 42	[98] **Tc** technetium 43	101 **Ru** ruthenium 44	103 **Rh** rhodium 45	106 **Pd** palladium 46	108 **Ag** silver 47	112 **Cd** cadmium 48	115 **In** indium 49	119 **Sn** tin 50	122 **Sb** antimony 51	128 **Te** tellurium 52	127 **I** iodine 53	131 **Xe** xenon 54
133 **Cs** caesium 55	137 **Ba** barium 56	139 **La*** lanthanum 57	178 **Hf** hafnium 72	181 **Ta** tantalum 73	184 **W** tungsten 74	186 **Re** rhenium 75	190 **Os** osmium 76	192 **Ir** iridium 77	195 **Pt** platinum 78	197 **Au** gold 79	201 **Hg** mercury 80	204 **Tl** thallium 81	207 **Pb** lead 82	209 **Bi** bismuth 83	[209] **Po** polonium 84	[210] **At** astatine 85	[222] **Rn** radon 86
[223] **Fr** francium 87	[226] **Ra** radium 88	[227] **Ac*** actinium 89	[261] **Rf** rutherfordium 104	[262] **Db** dubnium 105	[266] **Sg** seaborgium 106	[264] **Bh** bohrium 107	[277] **Hs** hassium 108	[268] **Mt** meitnerium 109	[271] **Ds** darmstadtium 110	[272] **Rg** roentgenium 111	[285] **Cn** copernicium 112	[286] **Uut** ununtrium 113	[289] **Fl** flerovium 114	[289] **Uup** ununpentium -15	[293] **Lv** livermorium 116	[294] **Uus** ununseptium 117	[294] **Uuo** ununoctium 118

*The lanthanides (atomic numbers 58–71) and the actinides (atomic numbers 90–103) have been omitted.
The relative atomic masses of copper and chlorine have not been rounded to the nearest whole number.

Physics Equations

force = mass × acceleration

kinetic energy = 0.5 × mass × (speed)^2

work done = force × distance (along the line of action of the force)

$$power = \frac{work\ done}{time}$$

$$efficiency = \frac{useful\ output\ energy\ transfer}{total\ input\ energy\ transfer}$$

weight = mass × gravitational field strength *(g)*

gravitational potential energy = mass × gravitational field strength *(g)* × height

force applied to a spring = spring constant × extension

moment of a force = force × distance (normal to direction of the force)

distance travelled = speed × time

$$acceleration = \frac{change\ in\ velocity}{time\ taken}$$

wave speed = frequency × wavelength

charge flow = current × time

potential difference = current × resistance

power = potential difference × current

power = (current)^2 × resistance

energy transferred = power × time

energy transferred = charge flow × potential difference

$$density = \frac{mass}{volume}$$

$$\text{pressure} = \frac{\text{force normal to a surface}}{\text{area of that surface}}$$

$$(\text{final velocity})^2 - (\text{initial velocity})^2 = 2 \times \text{acceleration} \times \text{distance}$$

change in thermal energy = mass × specific heat capacity × temperature change

thermal energy for a change of state = mass × specific latent heat

elastic potential energy = 0.5 × spring constant × (extension)^2

pressure × volume = constant (for a given mass of gas and at a constant temperature)

$$\text{efficiency} = \frac{\text{useful power output}}{\text{total power input}}$$

$$\text{period} = \frac{1}{\text{frequency}}$$

$$\text{magnification} = \frac{\text{image height}}{\text{object height}}$$

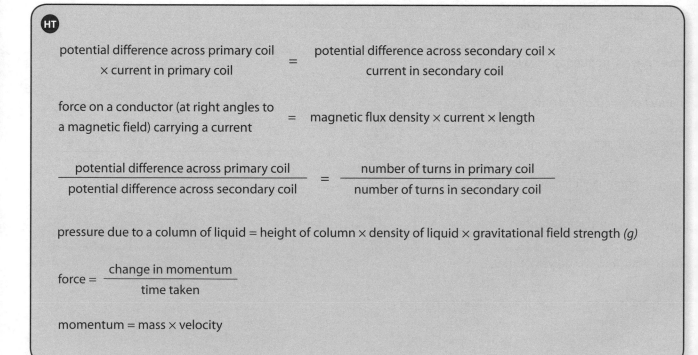

HT

$$\begin{array}{c}\text{potential difference across primary coil} \\ \times \text{ current in primary coil}\end{array} = \begin{array}{c}\text{potential difference across secondary coil} \times \\ \text{current in secondary coil}\end{array}$$

$$\begin{array}{c}\text{force on a conductor (at right angles to} \\ \text{a magnetic field) carrying a current}\end{array} = \text{magnetic flux density} \times \text{current} \times \text{length}$$

$$\frac{\text{potential difference across primary coil}}{\text{potential difference across secondary coil}} = \frac{\text{number of turns in primary coil}}{\text{number of turns in secondary coil}}$$

pressure due to a column of liquid = height of column × density of liquid × gravitational field strength (g)

$$\text{force} = \frac{\text{change in momentum}}{\text{time taken}}$$

momentum = mass × velocity

Index

The author and publisher are grateful to the copyright holders for permission to use quoted materials and images.

All images are © Shutterstock and © HarperCollins Publishers.

Every effort has been made to trace copyright holders and obtain their permission for the use of copyright material. The author and publisher will gladly receive information enabling them to rectify any error or omission in subsequent editions. All facts are correct at time of going to press.

Published by Letts Educational
An imprint of HarperCollins*Publishers*
1 London Bridge Street
London SE1 9GF

ISBN: 9780008276065

First published 2018

10 9 8 7 6 5 4 3 2 1

© HarperCollins*Publishers* Limited 2018

All rights reserved. No part of this publication may be reproduced, stored in a retrieval system, or transmitted, in any form or by any means, electronic, mechanical, photocopying, recording or otherwise, without the prior permission of Letts Educational.

British Library Cataloguing in Publication Data.

A CIP record of this book is available from the British Library.

Author: Dan Foulder
Commissioning Editors: Clare Souza and Kerry Ferguson
Editor/Project Manager: Katie Galloway
Cover Design: Amparo Barrera and Sarah Duxbury
Inside Concept Design: Ian Wrigley
Text Design and Layout: Nicola Lancashire at Rose & Thorn Creative Services, and Ian Wrigley
Production: Natalia Rebow
Printed and bound by Grafica Veneta S.p.A